Military Strategy

Military Strategy

The Politics and Technique of War

John Stone

BLOOMSBURY ACADEMIC
LONDON · NEW YORK · OXFORD · NEW DELHI · SYDNEY

BLOOMSBURY ACADEMIC
Bloomsbury Publishing Plc
50 Bedford Square, London, WC1B 3DP, UK
1385 Broadway, New York, NY 10018, USA

BLOOMSBURY, BLOOMSBURY ACADEMIC and the Diana logo are
trademarks of Bloomsbury Publishing Plc

First published in Great Britain 2011 by the Continuum International Publishing Group Ltd
Reprinted by Bloomsbury Academic 2013 (twice), 2015, 2016 (twice), 2018

A catalogue record for this book is available from the British Library.

A catalog record for this book is available from the Library of Congress.

ISBN: HB: 978-1-4411-6647-0
PB: 978-1-3501-0624-6
ePDF: 978-1-4411-8005-6
ePub: 978-1-4411-7294-5

Typeset by Newgen Imaging Systems Pvt Ltd, Chennai, India

To find out more about our authors and books visit
www.bloomsbury.com and sign up for our newsletters.

Contents

Preface

Yet another work on military strategy, in a market already crowded with related offerings? Surely an explanation is required – and here it is. My purpose in writing this book was to explore what I consider to be the fundamental challenge facing strategists, namely the selection of the objective in war. Readers of Clausewitz will be aware that this issue lies at the very centre of his theoretical engagement with *der Krieg*. Having deduced that all military activity ought, logically speaking, to aim at the disarmament of one's enemy as rapidly as possible, he then spent a great deal of time and effort endeavouring to explain why it is that such an objective is not always pursued in practice, and what the alternatives might be. The result of his efforts was the realization that warfare is a truly political activity: not only do political differences give rise to war, but they also shape the objectives that force is used to achieve in war. In this regard, Clausewitz's work prefigured our more recent concept of 'limited' war.

This, of course, is another way of saying that strategy is (or at least should be) a profoundly political activity; and in common with all such politicized endeavours, success demands the exercise of sound judgement. The use of armed force in order to bend an adversary to our will admits of no golden rules; there are a few general principles to be borne in mind, but the relative weight one attaches to each of them depends very much on the prevailing political context. It follows from this that one of the most valuable assets that a strategist can enjoy is a good understanding of his adversary's motivations for fighting, along with the strength with which they are held. Such an understanding provides the best possible basis for the exercise of judgement in strategic matters. By way of contrast, even a pronounced superiority in technical means is not nearly so valuable as a great many strategists have persisted in believing over the years. Not only is it remarkably difficult to sustain a truly significant margin of military-technical superiority in the face of a determined and resourceful enemy but, in the absence of a good understanding of this enemy, and the most efficiently applied force can be irrelevant or even inimical to one's own political goals. Our wars in both Iraq and Afghanistan illustrate this all too clearly: despite enjoying a huge margin of military-technical superiority at the beginning of both of these conflicts, the United States and its allies encountered great difficulties in wresting anything of political value from such a state of affairs. The inability of technical initiatives to substitute for a lack of understanding of one's adversary is, in fact, a key message of this book. It is by no means a new message but, having said that, the difficulties we have encountered in our efforts to fight a 'War on Terror' suggest it is one that needs to be repeated on a regular basis.

But to return to those strategy texts already sitting confidently on library shelves: do they not suffice to make yet another one redundant? That I have subsequently written this book is due to my belief that there remains a gap to be filled between what I consider the two main schools of writing about strategy that inform these texts. The first of these seeks to derive recommendations for strategic action via a process of logical deduction, which itself proceeds from a set of premises relating to the dynamics of strategic decision-making. As such it is instantiated by the game-theoretical approach, and by related attempts to theorize deterrence during the Cold War. Pleasingly parsimonious, and sophisticated in their own terms, such approaches nevertheless suffer from a failure to engage adequately with the challenge of translating theory (via the application of judgement) into practice in any given political context. To draw a somewhat waspish analogy, the strategic edifices constructed upon the foundations provided by this approach might be likened to the model cathedrals built from matchsticks that one occasionally encounters in exhibition halls: clever enough, but lacking the stamp of true art.

The second school is of the more traditionally historical variety. As such it is necessarily linked into the world of practice, although not infrequently at the cost of forgoing a solid basis in theory. Consequently, a great deal of this historical material does not really engage with strategy as a problem in the selection of objectives. Instead we read a lot more about how developments in military means have permitted war to be waged on an ever-grander scale – the assumption being that this is ultimately important for politics in some manner or other. Such accounts are, in other words, all too frequently underpinned by unproblematized assumptions of technical determinism. War is treated as a continuation of technique, as opposed to politics, by other means.

What I have tried to do in this book is to bring the most satisfactory aspects of these two approaches together by providing an explicit theoretical discussion of the dynamics of strategic decision-making, along with the manner in which political and technical considerations exercise an influence in this regard. I have then sought to use this discussion as a framework within which to explore certain historical instances of strategic practice that have occurred between the French Revolution and the present day. My goal, in other words, is explicitly theorized history.

Because strategic practice reflects to some degree a cumulative body of knowledge, I have found the chronological approach to be a valuable way of organizing my book. The results should not be seen as an attempt to write a comprehensive account of strategy during this period of two centuries, however. Rather, I have been concerned to do enough merely to explore the dynamics of military strategy under a variety of political and technical conditions.

As such, the research process underpinning this book involved the selection and interpretation of a fairly eclectic range of sources in the light of my interpretative framework. Wherever possible I have used primary sources, although a good many of them are well known enough to have been widely cited elsewhere. The originality of my book therefore lies in the manner in which the sources have been employed in order to illustrate the influence of a set of strategic dynamics that would seem to be consistent across time and space. Having said that, any synoptic work of this kind inevitably owes a great deal to those who have previously worked over the same ground, even if with rather different goals in mind. I have taken the view that much of this previous endeavour now constitutes generally accepted knowledge in the field. I have therefore elected to footnote only those claims that (to me at least) did not seem to be part of this accepted knowledge, along with specific quotations and certain instances of precise numerical information. I hope that in doing so I have not failed to acknowledge anybody's original contribution to this knowledge or our interpretation of it.

Likewise, so many other people have contributed, in one way or another, to this project that it would be both a very long job indeed to list them all, and tendentious to single out but a few. I shall therefore limit myself to saying thank you to them all. Finally, I should like to dedicate this book to my little daughter Jemima, in the hope that she may never have to experience the vicissitudes of war.

Introduction

In late March 2003 US forces invaded Iraq in the opening stages of an operation intended to topple Saddam Hussein from power. Operation 'Iraqi Freedom' was conceived as an important move in the wider 'War on Terror'. Removing Saddam was expected to transform Iraq from a 'rogue' state that might provide help to al-Qaeda into a democratic ally in the struggle against international terrorism. The resulting blitzkrieg provided yet another demonstration of US military-technical prowess: by the beginning of May regular Iraqi forces had been scattered or destroyed, Saddam was in hiding, and President George Bush Jr felt able to declare an end to 'major combat operations'.

Of course we now know that the month of May amounted to the end of nothing very much at all. On the contrary, it marked the beginning of what might be considered the real war in Iraq: a war that would see US troops struggling to hold together a shattered nation in the face of a vicious insurgency and mounting sectarian violence. Far from dealing al-Qaeda a body blow, the United States found itself drawn into a disaster of its own making. Unable to bring the violence to a rapid halt, and unwilling to leave Iraq as a new haven for terrorists, Washington faced a rising tide of casualties along with massive economic costs. Resources that might more profitably have been employed against terrorist threats elsewhere in the world were sucked into the quagmire and expended there. In consequence, the War on Terror became a much greater challenge than had previously been the case.

Many reasons have been advanced for the misfortunes that befell the United States in Iraq. At a fundamental level, however, US problems flowed from the Bush administration's seemingly unalloyed faith in the capacity of superior military technique to provide definitive solutions to complex political problems. Since terrorists could not readily be directly targeted with force, the next best option was to deprive them of bases and support by targeting those states that might be supplying such help. Unlike elusive terrorists, states are not difficult to find and attack: on the contrary they are exceedingly vulnerable to highly accurate strikes of the kind that the US military does best. Such a strategy, it was believed, would therefore play to US strengths and spare Washington from the messy and protracted politicking that less muscular approaches to tackling terrorism would have entailed. The danger that decapitated nation-states would not automatically grow new democratic heads, but would instead collapse in on themselves, quite possibly breeding yet more terrorists in the process, does not seem to have been seriously considered by senior figures in the Bush Administration who were prone to viewing force as a means of sweeping away obstacles to a natural progression of peoples towards liberal capitalism.[1] What mattered, it was believed, was the possession of suitably advanced military technique along with the will to use it when the opportunity presented itself.

While such views might well be considered more than a little naive, history offers many other instances of over-reliance on military-technical responses exacerbating problems when a more prudent regard for the political context might have yielded better results. Indeed Hans Morgenthau was once moved to observe that the United States tended to treat war as a 'self-sufficient, technical enterprise, to be won as quickly, as cheaply and as thoroughly as possible and divorced from the foreign policy that preceded and is to follow it'.[2] Moreover, if the United States is prone to unwarranted bouts of technophilia, it is certainly not alone in this regard. Striking an appropriate balance between the military and political dimensions of warfare is the job of *strategy*, and there are sound reasons for believing this to be a very demanding job indeed. Accurately gauging the scope available for the effective employment of armed force in any given political context demands that rarest of commodities: sound judgement. There are no hard and fast rules in this regard, and such matters must be worked out almost, as it were, through instinct. Little wonder, then, that a persistent tendency exists to seek certainty through superior military technique, to reduce warfare to an exercise in the efficient application of force, and to marginalize political considerations until such time as the shooting stops and a new peace must be built.

In writing this book I have been concerned to underline the point that reducing warfare to a military-technical enterprise is, under most circumstances, counter-productive; that formulations for the application of armed force must give due regard to the broader political context in which they are made. To be sure, politicizing warfare in this manner is a difficult, and indeed risky, undertaking; but in the absence of such efforts the use of force is all too likely to become counter-productive: force begets force and technique is set against technique, producing a spiral of violence whose ultimate costs can readily eclipse the political differences that brought about war in the first place.

In the chapters that follow, I have developed this basic observation with reference to various historical episodes of strategic practice over the past two centuries or so. My selection in this regard is not intended to provide a comprehensive coverage of technical modalities, national traditions, or the like. Naval strategy, for example, receives short shrift, while the second half of the book concentrates almost exclusively on certain strategic dilemmas faced by the United States since the period between the two world wars. Moreover, the accounts themselves are somewhat schematic: in each case I have focused on the manner in which strategy derived its character from the particular set of political and military-technical considerations that were prevalent at the time, and the extent to which the balance struck between them was successful. Extraneous issues have been omitted, even when they might be considered to have exercised an important influence on the outcome of the hostilities under consideration. My object is not to provide a well-rounded history, but to examine the formulation of strategy under various different political and technical conditions. Indeed, even as an exercise in strategic analysis the chapters that follow might be deemed rather narrow by some readers. On the manner in which ethical and legal considerations (to name but two examples) influence decisions on the application of force I have nothing to say. Additionally, my analysis rests heavily on the assumption of rationality in strategy, an assumption that is widely understood to be as problematic as it is necessary. About these and other neglected matters, I would once again say that they are not essential to my particular project and I have therefore set them aside.[3]

Having ruled a great many things out of bounds, let us now turn to a more detailed consideration of what remains. I propose to do this by engaging with three questions: what is strategy? how does strategy 'work'? and why is strategy difficult? Answering these questions will permit us to appreciate more clearly why effective strategy demands the application of sound judgement, and why in turn strategists have frequently been concerned to substitute such judgement with military-technical initiatives.

What is strategy?

For present purposes, I propose to define strategy as the instrumental link between military means and political ends. Strategy in other words is concerned with the process by which armed force is translated into intended political effects. Locating strategy at the interface between military means and political ends in this manner is something of a simplification, albeit one that serves a valuable purpose in terms of clarifying the scope of our enquiry.

By specifying *armed force* as the means available to strategy we avoid trespassing into the domain of grand strategy, an activity that is concerned with the application of the totality of national resources in the pursuit of political goals. A consideration of these many resources would lead us into areas such as diplomacy, economics and propaganda, thereby transcending the narrower concerns of this present work. While we cannot proceed without touching on such matters from time to time, it is armed force that constitutes our focus in this regard.

Moving to the other side of the equation, specifying the ends of strategy as *political* in nature permits us to delineate our area of interest from military operations and tactics. These latter two spheres of activity are really aspects of military *technique*, whose concern is with the efficient application of force. As such they are properly considered part of the means available to strategy, technique being what puts the 'armed' into armed force. We shall have more to say on the subject of technique later. In the meantime, let us consider the manner in which strategy provides the instrumental link between military means and political ends.

How does strategy 'work'?
At base, strategy involves the translation of political goals into one or more subordinate objectives that are amenable to the application of armed force. As such, the process is nicely summed up by Peter Paret, according to whom strategy 'is the use of armed force to achieve the military objectives and, by extension, the political purpose of the war'.[4] Formulating these subordinate objectives is, however, easier said than done. This is because the underlying decision-making process must take the response of a reasoning opponent into account. Our choice of objectives must, in other words, anticipate and forestall countervailing action by our adversary. With this point in mind Thomas Schelling characterized strategy as interdependent decision-making, because our optimum course of action depends on what we anticipate our adversary's response will be – a response that will in turn be conditioned by his expectation of how we shall respond to him.[5]

Proceeding deductively from a similar starting point, Clausewitz famously argued that our only logical course of action in war is to destroy our adversary's means of resistance as rapidly as possible. This is because the most dangerous response our adversary can possibly make is to destroy our means of resistance, thereby rendering us incapable of challenging his political goals. In order to forestall such a disastrous eventuality we must, therefore, disarm him at the earliest opportunity. To reject this necessity would only be to facilitate his efforts to disarm us – efforts he will be driven to make by the self-same logic that drives us to disarm him. In such a manner, the interdependent character of strategic decision-making encourages a rush to extremes: regardless of our political goals, our strategic objective must always be the rapid disarmament of our adversary.

However, Clausewitz went on to observe that strategic decisions are not solely a function of some disembodied logic: they are also conditioned by the value we place on victory, which is to say on achieving our political goals. More specifically, we do not want the costs of fighting to outweigh the benefits we associate with victory, for to permit this would be to make war a counter-productive activity. Thus from this broader political perspective, unrestrained efforts to disarm our adversary look distinctly less attractive because they are likely to prove disproportionately expensive in terms of blood and treasure. Even if we ultimately prevail, we shall be forced to endure the worst a desperate adversary can throw at us before he is finally disarmed.

In practice, therefore, cost-benefit calculations qualify the virtue associated with attempting to disarm our adversary – just as they qualify the virtue our adversary associates with attempting to disarm us. Consequently we tend to eschew such extreme efforts and, (reasonably) confident that our adversary will reciprocate in kind, settle on a rather more modest strategic objective. Such objectives typically involve destroying some increment of our adversary's means of resistance while holding the threat of additional destruction in reserve.[6] In such cases, the purpose of the exercise is to convince our adversary that continued resistance is futile, that he cannot hope to win, and thus by continuing to fight he will incur additional costs to no good purpose. We are trying in other words to *coerce* our adversary into submission, to undermine his resolve to fight rather than completely depriving him of the means necessary to do so. In all of this, therefore, our choice of strategic objective flows not from the most dangerous response our adversary can possibly make, but from his most probable response in light of the value he places on victory. It is exactly this conditioning of strategic choices by political considerations that Clausewitz had in mind when he characterized war as a continuation of politics by other means.[7]

Why is strategy difficult?

It is not particularly difficult to grasp the dynamics of strategic decision-making outlined above.[8] Problems do arise, however, when we move from general principles to their application in specific instances. It is one thing to claim that victory at acceptable cost is gained via politically conditioned strategic decisions that reflect the most probable, as opposed to the most dangerous possible, response available to our adversary. It is something else to establish precisely what that response will be in any given political context. Try as we might, we cannot determine exactly what value our adversary places on victory, and thus we cannot be certain of how much scope we enjoy for formulating strategic objectives that fall short of disarming him. We are haunted by the possibility that a deliberately restrained blow will fail to break our adversary's resolve; that moderation on our part will merely facilitate a forceful response intended to break *our* resolve or – worse still – a hammer blow designed to disarm us outright. We worry, in other words, that our efforts to win at bearable cost might well mean that we do not venture enough to avoid defeat.

This in turn means that judgement must step in to help us distinguish between the worst possible and most probable consequences of our decisions, and accept the latter as our warrant for action. According to Clausewitz, strategic decisions require 'skill in discriminating, by the tact of judgement among an infinite multitude of objects and relations, that which is the most important and decisive.'[9] More recently, Isaiah Berlin summed up the matter nicely when he described judgement as entailing

a capacity for integrating a vast amalgam of constantly changing, multicoloured, evanescent, perpetually overlapping data, too many, too swift, too intermingled to be caught and pinned down and labelled like so many individual butterflies. To integrate in this sense is to see the data . . . as elements in a single pattern, with their implications, to see them as symptoms of past and future possibilities, to see them pragmatically – that is, in terms of what you or others can or will do to them, and what they can and will do to others or to you.

Such a capacity, it need hardly be observed, is a 'gift': it is lacking to any pronounced degree in most of us, and nor can it be taught to those who do not possess it.[10] This, I want to suggest, is why strategy is difficult: it makes demands that most of us cannot routinely hope to meet. As Clausewitz observed, strategic decision-making creates a space for 'our own apprehensions and those of others, for objections and remonstrances, consequently also for unseasonable regrets' to run rampant with baleful results. Because everything 'must be conjectured and assumed, the convictions produced are less powerful', as a consequence of which we are plagued by 'bewildering doubts.'[11]

We can, of course, seek to minimize these 'bewildering doubts' by ensuring that our strategic decision-making is as well informed as possible about our adversary's motivations and intentions. The better we understand our adversary in this regard, the more traction our judgement will enjoy in respect to the formulation of strategic objectives. To be sure, there are limits to what we can achieve in this regard: we cannot look directly into our opponent's mind, but can only infer how he will act from what he claims he will do; what third parties claim he will do; what he has previously done in other (more or less) similar circumstances; and other such indirect sources. Moreover, while we are assiduously gathering and analysing all this information, the situation may change in ways that throw our emerging conclusions awry. Consequently we must take very serious strategic decisions on the back of evidence that is at best incomplete and even to some degree contradictory. Thus, despite the historically documented fact that even a little understanding can be very much better than none at all, it is perhaps not surprising that alternatives have been sought to the deeply uncomfortable business of exercising judgement under conditions of considerable uncertainty. Indeed, as we shall see next, it is the putative capacity of military technique to substitute comfortable certainty for 'bewildering doubts' that accounts for a great deal of its attractiveness for strategists. For as Colin Gray has observed, 'it is easier to theorize about new ways of prevailing than to speculate honestly and imaginatively about possible enemy initiatives and responses.'[12]

Technique, friction and speed

By *military technique* I mean the vast panoply of artefacts, practices and concepts that are dedicated to the efficient application of force. Understood in this inclusive sense, the term refers not only to weapons and ancillary equipments, but also to the trained personnel who operate them, along with the concepts for organizing them into combat, command and logistical units, and the tactical and operational concepts that govern their employment in war.[13] Technique can be tacit as well as explicit in character: it can refer to things that soldiers know how to do themselves but that they cannot adequately explain to others. If Napoleon was military technique incarnate, he was also technique inarticulate in the sense that he never succeeded in setting down a systematic account of his methods. Consequently the task of distilling a Napoleonic technique from the Emperor's military exploits fell to Jomini, whose *Summary of the Art of War* became a standard text for nineteenth-century soldiers.[14] But whether or not a general's 'art' is explicitly articulated, its function is the efficient application of force and it therefore falls within the province of military technique.

The value of technique resides in its capacity for ameliorating the obstacles to efficient military action that Clausewitz famously brigaded together under the label of *friction*. Included under this term are the effects of uncertainty and chance along with human moral and physical failings, all of which conspire to ensure that military action always falls well short of perfection.[15] Not only must force be applied without full knowledge of an enemy's intentions, but it is also frequently unclear where the enemy's key means of resistance are to be found. Additionally, force is never applied in the absence of chance events such as the onset of adverse weather conditions or the failure of important equipment. Moreover, soldiers are prone to make serious mistakes if they are tired or frightened as a result of exposure to combat. Consequently, the strategic objectives we set for our forces are seldom readily realized in practice. Outcomes therefore take longer to achieve, which is an extremely important consideration given that every delay represents an opportunity for our adversary to act in furtherance of his own objectives.

Viewed from this perspective, the important consequence of pitting technique against the influence of friction is that it facilitates the compression of military activity in time. Technical developments do not permit us to achieve more per se. Rather they permit us to achieve any given strategic objective more rapidly, and thus at less cost to ourselves. Schelling once observed that we do not require nuclear weapons in order to commit genocide against a defenceless opponent. In principle at least, ice picks will suffice. The job will take longer to achieve without nuclear weapons, but the result is assured so long as our sense of purpose does not fail.[16] Of course, should our opponent acquire ice picks of his own, the time required to rid the world of him becomes a more important consideration: the longer we take, the more opportunity he will have to kill some of us during the process. Were we to respond by eschewing ice picks in favour of nuclear weapons, we would once again be in a position to annihilate our enemy without being hurt in return. The massive power of nuclear weapons means that he could be killed off before being able to turn his ice picks to account.

The link between technique and speed therefore takes us back to the balance between costs and benefits that governs our choice of strategic objectives. And in doing so it explains a great deal of the importance that has historically been attached to the pursuit of military-technical superiority. The greater our superiority, the more rapidly we can achieve our strategic objectives; and the more rapidly we achieve our objectives the less opportunity our enemy will have to pursue his objectives to our detriment. Indeed, if our technical superiority is sufficiently great, we should be able to operate so rapidly as to preclude any serious resistance on our adversary's part. He will always be reacting to our blows and thus will never be able to mount any of his own. In principle at least, therefore, superior technique banishes the requirement to exercise judgement in our choice of strategic objectives. Because the costs associated with disarming a technically inferior adversary should be relatively light, it makes sense to pursue this objective rather than run the risks associated with voluntarily restraining our use of force in the expectation (but not the certainty) that our adversary will reciprocate in kind.

Flies in the ointment

In practice, however, efforts to substitute technical superiority for political restraint rarely produce such desirable results. An important reason for this is that it can be remarkably difficult to sustain a margin of technical superiority that is wide enough for us to disarm our adversary without incurring heavy costs in the process. Indeed, whatever technical gains we make are likely to be offset by countervailing developments among actual or potential adversaries. As Edward Luttwak has observed, conspicuously successful military-technical developments tend to galvanize equally successful countermeasures. In such a manner, a pronounced technical advantage can be rapidly whittled away to the extent that continued reliance on it becomes positively counter-productive.[17] One consequence of this is that the quest for technical superiority frequently catalyses qualitative competition of a dysfunctional kind. During the Cold War, for example, such behaviour proved extremely costly in economic terms while producing weapons of such disproportionate power, in relation to any political differences that might conceivably have led to their use, as to eclipse any notion of a meaningful technical advantage. To be sure, the United States emerged from the Cold War as the world's foremost military power; but against this must be set the observation that technical competition is not the sole province of so-called 'advanced' strategic actors. Adopting a broad understanding of technique – one that encompasses such prosaic matters as organization and tactics as much as the latest guided weapons that fall off the bottom of stealth aircraft or drones – reminds us that even supposedly unsophisticated strategic actors can engage in meaningful military-technical competition. Imaginatively employed, the techniques of the guerrilla and the terrorist can work very efficiently in the face of the most advanced conventional forces. Such techniques may not suffice to disarm the United States, but recent events demonstrate that they can be efficient enough to prevent the United States from disarming its adversaries without risking costs that outweigh the value placed on victory. Even modest levels of technique can, in other words, serve to facilitate strategies of coercion.

The noiseless harmony of the whole

All this suggests that even the most technically sophisticated strategic actors will likely find it difficult to disarm an adversary without suffering undue costs in the process. Under most circumstances, therefore, they would be better advised to exercise restraint over the strategic objectives they set for themselves in time of war, which in turn implies an important role for the application of judgement in such matters. This at least was the view of Clausewitz, who many years ago acknowledged the role played by technique in putting the 'armed' into armed force. 'Violence', he observed, 'arms itself with the inventions of Art and Science in order to contend against violence.' But having acknowledged this point, he was also concerned to emphasize that effective strategy rested on much more than technical matters.

> A Prince or General who knows exactly how to organise his War according to his object and means, who does neither too

little nor too much, gives by that the greatest proof of his genius. But the effects of this talent are exhibited not so much by the invention of new modes of action, which might strike the eye immediately, as in the successful final result of the whole. It is the exact fulfilment of silent suppositions, it is the noiseless harmony of the whole action which we should admire, and which only makes itself known in the total result.[18]

For Clausewitz, therefore, technique (new modes of action) is not without importance. But this importance derives from the manner of its use in any given context. Technique, in other words, is not a substitute for judgement, but an additional consideration requiring the application of judgement in the formulation of strategy. As we shall discover in the following chapters, his appreciation of these matters was entirely correct.

Chapter 1 shows how the French Revolution destroyed the existing political consensus in Europe, thereby creating a space for military technique to exercise a far more important influence over the fortunes of states than had previously been the case. An important consequence was Napoleon's rise to political power on the back of his superior generalship. Emperor Napoleon lacked legitimacy in the eyes of his monarchical neighbours, but as long as he kept his political ambitions in check his excellent military technique would certainly have deterred efforts to overthrow him: the risks associated with fighting him would have outweighed the benefits associated with victory. In the event, however, Napoleon's ambition got the better of him. His quest for European hegemony galvanized opposition that even his superlative generalship could not cope with. In consequence, he was ultimately disarmed and toppled from power. Brilliant general though he was, Napoleon proved a poor strategist in the sense that he lacked the judgement necessary to discern the political limits beyond which his technique would be insufficient to sustain him in power.

Chapter 2 examines Prussian and then German strategy during the wars of unification and the run up to the First World War.

It shows how prime minister Otto von Bismarck successfully subordinated the action of Prussia's technically superlative army to wider political considerations, thereby preventing war from becoming a counter-productive activity. After Bismarck's fall from grace, however, Kaiser Wilhelm II's disastrous foreign politicking encouraged the formation of a powerful Franco-Russian alliance against Germany, thereby leaving her reliant for her security on the technical efficiency of her army. It was this efficiency that chief of staff Alfred von Schlieffen sought to bring to the highest possible pitch so as to permit a lightning offensive designed to disarm her enemies at the outbreak of war. Ultimately, however, German technique was not equal to the task and her 1914 offensive lost momentum under the influence of friction, in the process locking her into a prolonged struggle of unprecedented costliness. No amount of technical preparation, it emerged, could compensate German strategy for the deficit in judgement consequent on Bismarck's departure.

Chapter 3 shows how the costs of modern warfare, as exemplified by the lengthy struggle of 1914–18, were considered to have dramatically undermined war's political instrumentality. One consequence of this was a lively debate about the capacity of advanced military technique – in the form of mechanized ground and air forces – to restore this instrumentality by defeating an adversary extremely rapidly, and thus at correspondingly low cost. Detractors from this idea correctly argued that technical initiatives would galvanize countervailing developments among potential adversaries, which would restore the balance of technique and preclude rapid victories. Against this background, Basil Liddell Hart's attempts to identify a strategy capable of extracting political leverage from the costs of modern warfare – of seeking to 'punish' enemy forces while deliberately foregoing costly efforts to destroy them outright – stand out as particularly innovative. It subsequently emerged that such a restrained strategy would not have served to defy Hitler, whose sensitivity to costs was not what Liddell Hart believed, the latter's judgement on this point being awry. Subsequently, Liddell Hart's basic conception of strategy would seem better suited to exploiting the destructive potential of nuclear weapons, although it was US strategic concepts that came to the fore after 1945.

Chapter 4 considers US strategy during the decades 1941–1961. It shows how military technique was embraced in an effort to avoid the heavy costs associated with fighting wars to the bitter end in support of liberal goals – a problem highlighted by the American Civil War and the First World War. To this end the United States fielded sophisticated mechanized ground and air forces, before pioneering the development of atomic weapons. In no case, however, did technique live up to the expectations placed on it. Countervailing German developments exacted a heavy price during the Second World War, while the Soviet acquisition of atomic weapons levelled the playing field in this regard too. Against this backdrop, arguments emerged for moderating US war aims with regard to the Soviet Union. If the United States could not hope to destroy its ideological enemy at bearable cost to itself, then it must remain content to 'contain' Soviet expansion via correspondingly limited acts of force. Such arguments were, however, rejected by the Eisenhower Administration amidst concerns that long-term preparation for such 'limited' wars might ultimately militarize US society and undermine its economy. In its place, Eisenhower opted for yet another technical fix in the form of nuclear weapons, which dominated US strategy until the advent of the Kennedy Administration in 1961.

Chapter 5 examines the development of US strategy for nuclear war from its inception under the Eisenhower Administration. At its heart lay the idea that Soviet aggression could be deterred by the threat of nuclear retaliation. The key problem was that the Soviets soon possessed a large nuclear arsenal of their own, with the result that war would be a mutually devastating event. Washington would therefore be reluctant to unleash such a war, most particularly in response to local aggression directed against one of its many far-flung allies. Here then was a potential 'response gap' that might be exploited by the Soviets – a gap that many considered to require forces and strategies for waging a more limited form of war. Eisenhower continued to reject such initiatives, however, convinced that the possibilities for exercising political control over war with the Soviet Union were slight. Any small war would likely escalate into a big one, and thus the only way to prevent such a catastrophe was to avoid war altogether. Attitudes changed under the Kennedy Administration, which initially looked more favourably at the possibility of exerting political control over the conduct of nuclear war. In the event, however, it proved impossible to devise a strategy that offered any reasonable prospect of maintaining such control. Thus although the project was never formally abandoned, and was indeed reinvigorated from time to time, the Cold War ended without the challenge having been solved. In hindsight it appears that Eisenhower's judgement on the question was sounder than that of his critics. The only meaningful control that politics could exert over nuclear warfare was to prevent it from happening in the first place.

Chapter 6 examines the development of US strategy for conventional war during the second half of the twentieth century. Although prospects for imposing political control over the conduct of such wars appeared relatively promising, the experience of defeat at the hands of North Vietnam highlighted the dangers associated with exercising restraint in the application of force. This encouraged the view that conventional war should not be restrained, the only sensible objective being to disarm one's adversary as rapidly as possible. Thereafter, strategy was viewed as a technical activity focused on the efficient application of force – a position that was encouraged by the spectacularly rapid victory against Iraq in 1991, which was commonly attributed to the capacity of advanced 'information technologies' to dispel the influence of friction. Behind the scenes, however, victory still turned on Washington's capacity to prevent the application of force from becoming a politically counter-productive act. This was more obviously the case during the Kosovo War of 1999 during which technically sophisticated US forces were subjected to intensive political control in a manner reminiscent of Vietnam. Ultimately, the experience of conventional war-fighting during the second half of the century confirms the point that strategic objectives should always reflect the political context in which war is conducted, and that setting these objectives requires an exercise in judgement that cannot be substituted with superior military technique.

Chapter 7 returns us to the issues raised at the beginning of this introductory chapter. The 'War on Terror' was launched amidst misplaced confidence that the problem of targeting terrorist organizations with force would be circumvented by the capacity of advanced military technique to re-engineer the world into one that would be inimical to the terrorists' values, in the process exposing them to direct attack. The fact that this confidence was based on a spectacular error of judgement has been revealed by the course of events in both Afghanistan and Iraq, where the highly efficient application of force has caused more problems than it has solved. Within these local contexts, the emergence of counterinsurgency strategies is testament to a belated realization that military-technical virtuosity must be subordinated to political considerations if it is to produce desirable results. This represents welcome progress, although it is progress on only a limited front. Counterinsurgency has provided an emergency response to local disasters of Washington's own making – disasters that policy makers will henceforth fall over themselves to avoid multiplying. Regime change, in other words, is off the political agenda for the foreseeable future. Meanwhile, on the global stage terrorism continues to present challenges whose satisfactory resolution demands the application of force. History suggests that these acts of force must be rather more than mere exercises in technique: they must be subordinated to wider political considerations via a decision-making process that will always make heavy demands on the judgement of those involved. Good strategy rests on sound judgement. We neglect this lesson at our peril.

1 The French Revolution and Napoleon

No other year is more important for the history of Western strategy than 1789. Prior to the French Revolution, warfare in Europe had been distinctly limited in character. Wars were frequently cautious affairs characterized by the manoeuvre of armies rather than their whole-hearted commitment to battle, and were terminated more by mutual consent than by the complete defeat of one or other of the belligerents. In the wake of the Revolution the character of warfare underwent a profound change. Manoeuvre gave way to battle as the defining activity of military campaigns. War-termination lost its consensual aspect, and was now imposed on the losing side via a process of more or less comprehensive military destruction. As we shall shortly see, the root cause of this change was political in nature. While developments in military technique were certainly improving the efficiency of armies during the eighteenth century, such developments did not, in and of themselves, produce these dramatic changes in the character of warfare. Rather, it was the removal of political restraints on the formulation of strategic objectives following the Revolution that was most important in this regard. Under such circumstances the fortunes of France and its adversaries rested on the size and efficiency of their armed forces, in what amounted to a struggle for existential stakes. The result was a quarter century of bloody warfare before the enemies of France finally succeeded in imposing a new political order in Europe.

The eighteenth century

The conduct of war before 1789 was restrained in character chiefly because bitter ideological divisions were absent from European politics. States certainly went to war with each other on a frequent basis. Crucially, however, they also accepted the concept of dynastic legitimacy and the restraints this imposed on foreign policy. Generally speaking, a monarch's right to rule was still regarded as divinely sanctioned and therefore inviolate. In consequence, wars were not conducted for political goals of a kind that we would today term 'regime change'. Instead, monarchs typically resorted to war for the rather more modest purpose of territorial aggrandizement. Land remained the key form of wealth in early-modern Europe, and the question of who owned what was therefore of great importance to dynastic fortunes. Important as the question was, however, it was frequently difficult to answer definitively. The complex webs of intermarriage that had bound the European dynasties together over many centuries could frequently give rise to competing claims over territory, not least in the event of a succession crisis. In cases where dynastic claims collided, war was viewed as an acceptable option for resolving the matter.

There were of course always exceptions to the rule, with Frederick II of Prussia being a conspicuous example in this regard. Frederick's dynastic claims over Silesia and Saxony were tenuous, to say the least, and his seizure of them by force (in 1740 and 1756 respectively) provoked widespread outrage. Nevertheless, the exceptional character of these annexations was not mirrored in the response it elicited from the European powers that went to war against the Prussian 'rogue' state. Frederick's transgressions notwithstanding, the Hohenzollern dynasty's legitimacy could not be ignored by other right-thinking monarchs. Thus their political purpose in making war against him was not to overthrow him but merely to deprive him of his ill-gotten acquisitions and to cut him down to size.[1]

Poland, on the other hand, *did* disappear from the political map in 1795, a victim of partition by her more powerful neighbours. Although widely condemned at the time, partition was more readily conceivable in this particular case, however, because the Polish king had long been elected by a nobility who represented the real power in the land. As such, there was no dynastic legitimacy to preserve in the strict sense of the word. Thus when Stanislaw August Poniatowski (r. 1764–95) unexpectedly championed the introduction of a constitution designed to dilute the power of the nobility, they were quick to invite foreign intervention. Poland's neighbours were happy to oblige. Not only was there territory to be had, but progressive constitutions were decidedly worrisome in the immediate wake of the French Revolution.

Although exceptions such as these qualify the general rule, it should be remembered that Frederick's exploits and Poland's sad fate stand out precisely because they were exceptional cases. They were much commented on at the time and subsequently. In a sense, therefore, their prominence in the historical record reinforces the point that eighteenth-century wars were normally fought for limited political objectives.

Moving, now, to the other side of the ends–means equation, the slender character of the military resources available to eighteenth-century states also exerted an important restraining influence on warfare. In an age before mass politics, when the rarefied dynastic concerns of monarchs did little or nothing to capture the imagination of their subjects, the latter had either to be paid from finite treasuries or coerced into serving as soldiers. Casualties were correspondingly difficult and expensive to replace, which was a serious consideration when a few hours of battle routinely killed or wounded around one-third of the soldiers involved.[2] This made armies into precious commodities that were not to be risked lightly.

This combination of limited political objectives and precious military means encouraged a prudent form of strategizing that held no place for the single-minded pursuit of battle. Indeed, from a strategic perspective, the risks associated with resorting to battle tended to loom disproportionately large when set against the less-than-vital character of the territorial disputes that gave rise to them. This in turn encouraged the view that the epitome of good generalship consisted not in falling upon one's enemy at the first opportunity, but in artful manoeuvre designed to place him in a position from which he must either give battle on disadvantageous terms or retreat. Given the limited political stakes involved, an enemy who was placed at such a disadvantage was likely to retreat rather than fight, thereby ceding part of the territory under dispute. Considerations of this type led Maurice de Saxe to declare that

> I do not favor pitched battles, especially at the beginning of a war, and I am convinced that a skillful general could make war all his life without being forced into one. . . . I do not mean to say by this that when an opportunity occurs to crush the enemy that he should not be attacked, nor that advantage should not be taken of his mistakes. But I do mean that war can be made without leaving anything to chance. And this is the highest point of perfection and skill in a general.[3]

For his part, Frederick II greatly admired de Saxe but nevertheless departed from him on the matter of battle's necessity. 'War' he declared, 'is decided only by battles, and it is not finished except by them.' But Frederick, as we have already observed, was an exception in all such matters. Moreover, even he remained circumspect on the question of when to give battle, 'which should be done opportunely and with all the advantages on your side.'[4]

Far from being the epitome of military art, therefore, the precipitate resort to battle was typically regarded as a symptom of poor generalship. It might occasionally prove possible to punish an adversary's mistakes in the most severe manner, and battle might otherwise prove necessary in order to terminate a war outright. But under other circumstances a good general should find battle neither necessary nor desirable, and should seek to profit from the fact that his adversary would likewise consider it neither necessary nor desirable. Strategy, in other words, displayed a decidedly coercive character, the object of the exercise being to manoeuvre one's opponent into a situation where the acceptance of battle would most likely result in disproportionate costs on his part. As such it represented an appeal to the power of rational cost-benefit calculation, and in this regard faithfully echoed the prevailing Enlightenment concern with subordinating human affairs to the dictates of reason.

Developments in military technique during this period can likewise readily be understood within this Enlightenment framework. For example, the organization of armies into independent 'divisions' helped make the whole easier to manoeuvre than had previously been the case, while the introduction of standardized weapon designs simplified problems of maintenance and supply. Similarly, the burgeoning literature on the 'art' of war (of which de Saxe's work was an early example) contained various attempts to formalize the operational aspects of manoeuvre, rendering it more certain in its effects *vis-à-vis* one's adversary. In short, the thrust of military innovation was intended to rescue warfare from the 'bloody and conjectural' status that Voltaire is said to have assigned it.[5] The idea was to render war less susceptible to the influence of friction – to permit its being 'made without leaving anything to chance' – and thereby creating a more efficient instrument for the settlement of political disputes. Technical developments of this kind were not, therefore, intended to produce dramatic increases in the scale and scope of the violence associated with war. Rather, their desired effect was to rationalize the conduct of war within the restrictive framework of limited political objectives occasioned by dynastic disputes.

In the course of events, however, military technique would play a far more important role in shaping the character of warfare than had hitherto been imagined. For with the French Revolution came the abandonment of traditional political restraints on the formulation of strategic objectives. Under such conditions, the conduct of war was destined to be limited only by the resources available to the belligerent states along with their ability to employ them efficiently. If in the process warfare became somewhat less conjectural in its conduct, it also became far bloodier in its results.

The French Revolution

The origins and course of the French Revolution need not detain us here. What is important for our purposes is that it resulted in the overthrow of the Bourbon monarchy, and its replacement with a republican system of government. This development was highly significant because French republicanism derived its sovereignty from the people, and therefore recognized nothing sacred in the institution of monarchy. As one of the chief ideologues of the Revolution, the Abbé de Sieyès, put it: 'The nation exists before everything, it is the origin of everything. Its will is always legal, it is the law itself. Before it and above it there is only *natural* law.'[6] One can hardly imagine a more damning indictment of monarchy, an institution that was commonly understood to operate in exactly that space between nation and natural law that Sieyès was now claiming did not exist. Indeed it was for this reason that the great conservative thinker, Edmund Burke, felt moved to characterize the revolutionary wars as a struggle involving

> the partizans of the antient, civil, moral, and political order of Europe against a sect of fanatical and ambitious atheists which means to change them all. It is not France extending a foreign empire over other nations: it is a sect aiming at universal empire.

From this he concluded that 'this new system of robbery in France, cannot be rendered safe by any art; that it *must* be destroyed, or that it will destroy all Europe.'[7] Under such conditions the limited political aims that had previously reflected the restrained nature of dynastic squabbles did indeed give way to extreme efforts intended to erase the opposing ideology – a development that would in time exert an equally extreme influence over the character of strategy.

War conducted for existential stakes could only be won by one side or the other being rendered defenceless. Such conditions provided no occasion for politics to exert a restraining effect on the formulation of strategic objectives, which therefore extended to destroying the enemy's armed forces in battle. To seek anything less was merely to leave in place a mortal threat to oneself, and thus the battle of annihilation became the necessary goal of strategy rather than the recourse of poor or unlucky generals. Thus, whereas the wars of Frederick II resulted in just 12 major battles (involving a total of at least 100,000 soldiers), the wars consequent on the French Revolution and Napoleon's rise to power resulted in no fewer than 49 such battles.[8]

Not surprisingly, the logic of the new situation took some little time to filter down to the level of strategic practice, especially among the enemies of France who still habitually thought in terms of manoeuvre and the dominance of territorial objectives, and who were still necessarily concerned to minimize the losses inflicted on their precious long-service armies. This, in turn, could produce a dangerous disjuncture between political and strategic goals, as was notably the case at Valmy in 1792.

The battle of Valmy was the result of a combined Austro-Prussian effort conducted with the express purpose of crushing the revolution and restoring Louis XVI to the throne of France. To this end, a powerful force of Austrian and Prussian troops was concentrated on the Rhine, under the command of the Duke of Brunswick. The troops were considered to be Europe's best — long-service professionals possessed of considerable experience. Moreover, Brunswick himself was commonly understood to be one of the greatest generals of the age. A veteran of the Seven Years' War, he instantiated the eighteenth-century model of skill at manoeuvre tempered with a prudent sense of caution.

To make matters even worse for the French, when Brunswick commenced his march westwards the best that could be interposed between him and Paris was a disorganized assemblage of levies, bolstered by a rather more technically proficient contingent of artillery. Nevertheless, the French general in chief, Charles François Dumouriez, quickly grasped the grave threat posed to the Revolution and was determined on making a stand and forcing the issue: there would be no ceding ground along the route to Paris for him. This made it easier for Brunswick to envelop him, and matters would have gone badly indeed for Dumouriez but for the timely intervention of General François Christophe Kellerman whose hastily committed forces were used to extend the threatened French flank, thereby producing a head-on encounter with the Austro-Prussian army. At this point Brunswick's professionally ingrained sense of caution came to the fore. Having failed to dislodge his adversary by dint of manoeuvre, he was faced with the prospect of attempting to destroy him under less than ideal conditions. He subsequently essayed a couple of desultory frontal assaults that failed to break the French line and that cost him some hundreds of casualties. Had he continued in this manner he might well have succeeded in destroying the opposition. The cost to his valuable army would likely have been substantial, but the route to Paris would have been open and the revolution doomed. Brunswick's heart was not in the fight, however. Having been repulsed, he preferred to preserve his command for another day rather than risk great losses, even though these might well have been considered justified by the political stakes. '*Hier schlagen wir nicht*' (We will not fight here), he observed as he retired from the field in good order, thereby providing the revolution with an opportunity to draw breath and fight on under more favourable conditions.[9]

In the event, it was the French who first realized the logical consequences of the new political conditions as they pertained to the formulation of strategic objectives. Indeed, it was in this context that Lazare Carnot famously exhorted his generals to conduct operations intended to destroy the enemy's armed forces.

It is evident [he claimed] that we cannot win the war in this campaign without major battles because, through lesser operations, we would succeed only in destroying part of the enemy army, which would retain the means to attack us again the following year, and thus prolong the state of hostilities. We must, therefore, mount a most vigorous offensive campaign.[10]

Not surprisingly, this new emphasis on battle and destruction proved very expensive in terms of casualties. On the other hand, the popular character of the new Republic enabled it to raise unprecedented numbers of citizen soldiers from a (more or less) enthusiastic manpower base via mass conscription. Emergency legislation in 1793 introduced a *levée en masse*, as a result of which the size of the French army approached the three quarters of a million mark. Thereafter a more carefully organized system was developed in order to place conscription on a more sustainable basis. So long as the necessary manpower could be found, therefore, the French war effort would be constrained only by the ability of its generals to achieve the kind of crushing military victories on which the Republic's survival was considered to rest.

The Republic's adoption of a battle-seeking strategy sustained by mass conscription came as a shock to the enemies of France. And yet for all that, experienced generalship – a clear sense of what was operationally feasible in the face of friction's confounding influence – could still best an adversary whose chief advantage lay in a firm purpose bolstered by insensitivity to the costs associated with bold action. Superior technique could, in other words, still prevail over numbers and enthusiasm, as was notably demonstrated by Archduke Charles of Austria who, despite his cautious style, succeeded in inflicting a series of defeats on the French during the 1790s. Thus the new revolutionary armies were by no means everywhere successful despite the fact that, when they failed, their unfortunate generals frequently paid the heaviest of prices. Indeed, during the desperate years of 1792–93 no fewer than 84 of them were executed for disappointing their political masters back in Paris.[11] And then, at the other end of the scale came Napoleon Bonaparte, of whom it might justly be said that the Republic paid with *its* life for his victories.

The most conspicuously successful of the new French generals, Bonaparte was soon to become famous for his ability to overcome the effects of friction, to capture the essentials of the operational situation facing him and to commit his forces to battle under highly favourable circumstances. Jomini subsequently encapsulated the process as follows:

> To determine at a glance the possibilities offered by the various zones in a theatre of war; to mass his forces against the zone that offered the best prospects for success; to neglect nothing in his efforts to ascertain the enemy's approximate location; then to strike like lightning against the centre of his army were it extended, or otherwise against the flank that would lead him most directly to its lines of communication, to envelop it, cut it off, set it to flight, pursue it remorselessly, scattering it in all directions; to cease only after having destroyed or dispersed it: such was the basis on which Napoleon preferred to conduct his early campaigns.[12]

In consequence of his capacity for managing all this activity so efficiently, Bonaparte was destined to win a string of spectacular victories that would elevate him to the level of military genius and would endow his technique with exemplary status. And yet for all his military virtues, he certainly did not constitute a reliable instrument of state. On the contrary, he exploited the victories he achieved in the name of the Republic to pave his own way to power at the expense of his political masters. As early as the Italian campaign of 1796–97 he had revealed a disconcerting tendency to pursue his own political line by unilaterally establishing a series of satellite republics in the north of the peninsula and partitioning Venice. Thereafter, an expedition to Egypt briefly kept him at a safe distance from Paris, but upon returning in 1799 he quickly became involved with Sieyès in plotting a coup against his political masters. The incumbent Directory was struggling to cope with counter-revolutionary unrest at home and war abroad, and Sieyès now intended to replace it with a more effective alternative. In this regard he appears to have viewed his co-conspirator's military escapades as a valuable asset. The popular acclaim they had generated meant that Bonaparte would be seen as the acceptable face of military muscle, should a show of force be required to bring the coup to fruition. But if Sieyès was hoping to contain 'his' general's influence over subsequent events he was to be sadly disappointed. Although the coup was successful, in its aftermath Bonaparte showed considerable skill in out-manoeuvring the former abbé and having himself appointed as First Consul. Having arrogated power to himself in this manner, he then effectively consolidated his position in the eyes of the French people by defeating the Austrians in a dramatic battle at Marengo.[13] For the most part, therefore, it was his superior exercise of military technique that had brought Bonaparte to political power, and that would be what sustained him there – at least in so far as it did not fail him.

Emperor Napoleon and his wars

Napoleon's consolidation of his grip on political power in France created hopes abroad that he might ultimately be integrated into the polite society of European states. Certainly there were aspects of his behaviour that suggested this might prove possible in the long run. In 1804 he crowned himself Emperor of France with a view to founding his own dynasty. Thereafter he established a lavish court along pre-Revolutionary lines and created a new nobility from among his friends and relations. While such moves were viewed as the height of presumption abroad, they did suggest that Napoleon was no doctrinaire republican dedicated to the destruction of monarchy, and, for their part, his adversaries inclined towards conferring a de facto legitimacy on his regime. Ultimately, however, any notions of 'civilizing' Napoleon by binding him into the dynastic system came to nothing. Although he certainly adopted the trappings of monarchy and aped its ways, he was never willing to subscribe to the legitimacy of the institution and the restrictions this would have placed on his own ambitions, which were essentially unbounded. According to Napoleon's personal secretary:

He always considered war and conquest as the most noble and inexhaustible source of that glory, which was the constant object of his desire. He revolted at the idea of languishing in idleness at Paris, while fresh laurels were growing for him in distant climes. His imagination inscribed, in anticipation, his name on those gigantic monuments, which alone, perhaps, of all the creations of man, have the character of eternity.[14]

This was not the portrait of a man who would permit himself to be captured and constrained by the political orthodoxies of his day. Thus although he ultimately consented to marry Archduchess Marie-Louise, the daughter of Francis I of Austria, it was with his boot on her father's neck in the wake of war. Indeed, the scale of Napoleon's personal project – his quest for a unified Europe under his own control – meant that war was the only instrument available to him, as none of the monarchs standing in his way would acquiesce in his will without bitter resistance. Thus when he practised politics it was to enter into militarily advantageous alliances of a temporary nature; if he proved magnanimous in victory it was because he recognized the difficulties associated with bringing vast new territories and populations under direct control. A defeated monarch might therefore retain his throne if he proved willing to subordinate his own policies and resources to Napoleon's grand project. A self-declared adversary of feudalism, the new emperor was perfectly willing to adopt the practice when it suited his own purposes.

Reducing monarchs to the humble status of vassals demanded as a preliminary that their means of resistance – their armies – be destroyed. This in turn mandated a continuation of the battle-seeking strategy that had emerged in the wake of the Revolution, with all the associated manpower implications that this involved. By skilfully linking his personal ambitions to the fortunes of the French nation, Napoleon was able to continue the process of mass conscription that had been pioneered in the Republic. And as his empire grew so he was able to call on his subject states to augment the French army's ranks, although the local discontent this engendered sometimes placed limits on this particular policy.

As has already been noted, Napoleon's superior military technique enabled him to employ his forces to dramatic effect, time and again bringing his enemies to battle under circumstances that permitted him to inflict crippling defeat on them. A complete list of these battles would make for tedious reading, but it is worthwhile highlighting a selection of those victories that brought him to the summit of his imperial power. In 1805 he defeated Austrian and Russian armies at Ulm and Austerlitz, forcing Francis I to seek peace on highly unfavourable terms. Austria was required to cede important territories in Italy and Germany, and to pay a substantial indemnity. The following year Prussia was soundly defeated at the twin battles of Jena and Auerstädt (in the process providing a young Clausewitz with his formative military experience). The resulting peace terms reduced the state to a shade of her former power by stripping her of half her territory and imposing a swingeing indemnity. Throughout all this Russia fought on, but finally met with serious defeat at Friedland in 1807, leaving Napoleon master of continental Europe.

Indeed by 1807 Napoleon had, by dint of his military prowess, risen to the height of his political power. With the exception of insular Britain, which remained secure behind its navy, all those who dared raise a hand against him had encountered severe defeat on the battlefield. In consequence his writ now ran eastwards from Brittany all the way to the Russian border. On the other hand, exactly because his political power was founded on the superior exercise of force it lacked legitimacy in the eyes of the conquered. Indeed the experience of defeat, followed by a harsh peace, had the effect of building up a great deal of resentment against Napoleon – a resentment that was only kept in check by fear. The result was that his empire could only be held together by force, and thus his rule was always likely to be challenged if and when he suffered a serious military reversal. As Napoleon himself observed:

Your sovereigns, born on the throne can be beaten twenty times and still return to their capitals. I cannot do that because

I am an upstart soldier. My domination will not be able to survive from the day I cease to be strong and consequently to be feared.[15]

Meanwhile his opponents bided their time and sought to learn from their misfortunes. Reforms pioneered by Archduke Charles in Austria and General Gerhard von Scharnhorst in Prussia were intended to generate both the military technique and the man-power necessary to put the pursuit of battle at the centre of their respective strategies, in preparation for the day when Napoleon could be given a dose of his own medicine.

Given the heroic nature of the Napoleonic project, a major setback was only a matter of time. The first one came along in 1808 when he tried to consolidate his grip on Spain by placing his brother, Joseph, on the throne. In doing so, he ignited a popular revolt that French forces struggled to suppress. Realizing that efforts by irregular forces to destroy the French army in battle were fruitless, the Spanish abandoned such a strategy in favour of what we today would call an insurgency: the *Guerrilla*. Strong bodies of enemy troops were deliberately avoided in preference for hit-and-run attacks against smaller, isolated groups. The *Guerrillas'* superior knowledge of the local country, combined with the support of its inhabitants, meant that they operated under a smaller burden of friction than did the less agile French, thus permitting them to remain elusive until such time as they chose to strike again on their own terms. In this they were aided by a British expeditionary force operating out of Portugal, whose presence made it hazardous for the French army to disperse and hunt down the insurgents. In consequence the French found themselves locked into a prolonged war of attrition that extracted heavy costs in terms of blood and treasure. This was the first major reversal of fortunes experienced by imperial France, and Napoleon's domestic popularity suffered accordingly; moreover, the fighting kept 200,000 of his best troops busily occupied when they were badly needed in Germany. Little wonder, then, that he was moved to characterize the peninsular war as his 'Spanish ulcer'.[16]

Indeed, it was developments in Spain that encouraged Austria to re-open hostilities with France in 1809. As it happened, the fact that many of Napoleon's most experienced troops were committed to operations on the Iberian peninsula did not prevent him from punishing Austria for its presumption. Archduke Charles succeeded in inflicting a rare reversal on the emperor at Aspern-Essling, but the tables were turned a few weeks later at Wagram where Napoleon well and truly trounced him. The resulting peace terms stripped Austria of yet more territory that included her Adriatic ports, and imposed on her another heavy indemnity. All this was intended to minimize the chances of Austria causing Napoleon any further problems in the future, an outcome that he subsequently sought to perpetuate by marrying the Archduchess Marie-Louise. In the event, however, Francis I would only remain compliant for as long as his new son-in-law could compel him to do so, and in 1812 Napoleon lost the means necessary to achieve this.

It was only a matter of time before political tensions with Russia came to a head in some form of a violent showdown. Russia's vast reservoir of manpower meant that her military potential was considerable indeed, while the increasingly restive Tsar Alexander I was no friend to the French emperor. On the other hand, the size of the country precluded anything approaching a comprehensive occupation effort. Napoleon therefore concluded that a buffer zone should be created between his empire and that of Alexander, and in 1812 he decided that the time was ripe to carve it out by force. The resulting invasion of Russia constituted his most ambitious undertaking and resulted in his greatest military disaster. The army that was assembled for the enterprise was so large as to generate its own burden of friction that even Napoleon (who was himself in poor health at the time) could not overcome. As his forces crawled eastwards towards Moscow they exhausted their overstretched supply structures, only to discover that retreating Russian forces had laid waste to the territory in their path.

Thus, as Charles Minard's 1869 graphic of the Russian campaign vividly illustrates, the French army was being rapidly whittled away almost from the moment it crossed the Nieman. Napoleon finally succeeded in bringing Russian forces to battle at Borodino, where he inflicted some 50,000 casualties on them. But despite suffering such appalling losses Alexander nevertheless refused to capitulate, choosing instead to cede Moscow while his forces regrouped to the east. Critically short on supplies, and denied winter quarters by the burning of Moscow, Napoleon found himself with little option but to retreat westwards. Russian forces followed in hot pursuit, harrying the now starving invaders to almost complete destruction. Of the 422,000 soldiers that originally comprised the French invasion force, only 10,000 made it back to French-occupied territory.[17] In such a manner did Napoleon's ill-fated adventure in Russia finally provide the opportunity for a general European uprising against his hegemony.

Brilliant general, poor strategist

For Jomini, there was something of the morality tale in Napoleon's meteoric career. 'One might say that he was sent into this world to instil caution in generals and heads of state alike: his victories are lessons in skill, activity and audacity; his disasters demonstrate what can flow from a want of prudence.'[18] Certainly the trajectory of Napoleon's fortunes demonstrated that even generalship of his calibre could provide no sustainable alternative to a prudent foreign policy. True enough, his superior generalship served him remarkably well. It helped bring him to political power and to make France the dominant state in Europe. Used with greater discretion it could thereafter have secured his position as self-made Emperor of the French; for whatever misgivings his monarchical neighbours may have entertained about the legitimacy of such political arrangements, the military risks associated with trying to topple Napoleon from his throne would certainly have been considered too grave to countenance. So long as he did not make too much trouble, therefore, he would have been left alone to focus on the ticklish problem of ensuring his succession.

In the event, however, Napoleon's unbridled political ambitions finally convinced his enemies that the costs of accommodating him outweighed the risks associated with striking back when the moment seemed propitious. That moment finally arrived in 1813. With the French army dramatically weakened by the massive casualties it had suffered the previous year, the powerful coalition of states that took to the field against France could now generate a numerical superiority that even Napoleon's generalship was unlikely to overcome. For his part Napoleon remained determined to fight, and managed to scrape together something approaching 180,000 soldiers with which to confront a combined Austrian, Prussian and Russian force of around 300,000 at Leipzig in late October. Napoleon succeeded in landing some powerful blows on his adversaries, but weight of numbers eventually told and he was ultimately bundled out of Germany with a casualty bill that he could no longer make good.

The way was now clear for an invasion of France in early 1814, but even at this late stage his enemies were willing to negotiate a cessation of hostilities that would have left Napoleon on his throne. Ever mindful of his capacity for engineering startling military victories from the most inauspicious circumstances, they yet remained nervous about pushing him too far. Only when it became clear that there was no hope of Napoleon compromising over power, that the only way to achieve a lasting peace was to topple him from his throne, did his enemies press on with the task of disarming him. With his army in tatters and a war-weary France unwilling to rise up in response to invasion, Napoleon now had no alternative but to abdicate.

From this perspective it would seem appropriate to characterize Napoleon as a poor strategist. The matter turns not simply on the fact that he was ultimately defeated, but on why he was defeated. Napoleon was certainly a superb military technician, and was not without considerable merits as a head of state. But it was in striking a sustainable balance between his generalship and his political ambition that he fell down. In short, he lacked the judgement necessary to discern the political limits beyond which the efficient application of force could reliably sustain him. In Napoleon's case this was not so much because he failed to understand his enemies; rather, it was because he demonstrated what Charles Esdaile has described as a 'reckless disregard for international sensibilities'.[19] He simply did not care about such matters, and in consequence set a course that emboldened a spirit of grim resistance among his monarchical 'cousins'. Napoleon was never reliably matched in the area of generalship, although Archduke Charles was very much a threat in this regard, and the *Guerrilla* constituted an alternative set of techniques for tackling otherwise unassailable regular forces. However, it was his own actions that helped to create the weight of numbers that offset this technical deficit and finally swamped him. Such is the fate of those who permit their technique to flatter their ambition, thereby blinding them to the political realities in which they operate.

2 Strategy in nineteenth-century Prussia and Germany

In the wake of Napoleon's final defeat at Waterloo, the conservative statesmen of Europe endeavoured to forge a new system of monarchical states whose interests would be sufficiently congruent to preclude a repetition of the highly destructive events of 1789–1815. In this, however, they were to be hindered by vigorous nationalist sentiment along with its accompanying doctrine of self-determination. The bitter struggle against both republican and Napoleonic France had done much to awaken nationalist ideas among its enemies. Moreover, while the experience of French conquest and rule had frequently been a bitter one, Napoleon had effectively demonstrated that viable alternatives existed to the feudal political structures of pre-revolutionary Europe. Thus it was that Napoleon himself was outlived by the disturbingly liberal idea that nations composed of sovereign peoples should form the basis of political life, and that anointed kings remained an obstacle to achieving this goal.

The political upheavals that culminated in the second French Revolution of 1848 ultimately convinced conservative Europe that such sentiments could not be stamped out, and that some degree of accommodation was therefore necessary in order to stave off disaster. It was in this context that heads of state began deliberately to espouse nationalist values and aspirations in an effort to undermine the allure of liberalism. If such values and aspirations could be realized under the auspices of a king or an emperor, it was reasoned, then calls for self-determination were more likely to fall on deaf ears. By and large, this approach was to prove remarkably successful in tying nation and state together.

Whereas nationalism had begun its political career as the enemy of divinely sanctioned monarchs, it was subsequently transformed into a new form of spiritual legitimacy for heads of state and their governments. Nowhere was this more obviously the case than in Prussia, where war played a conspicuous part in the process.

For its part, Prussia experienced sweeping political changes as a result of 1848. A shaken Friedrich Wilhelm IV retained his throne, but liberal-nationalist sentiment was not to be denied the establishment of a parliament that subsequently represented a serious challenge to absolute monarchical power. Indeed from mid-century, the Prussian prime minister Otto von Bismarck realized it would be folly to attempt the direct repression of such representative institutions. If the position of the monarchy was to be preserved over the long term, it could only be achieved by co-opting parliament to the task. In the words of Otto Pflanze: 'Bismarck was the political surgeon who amputated nationalism from liberalism', and who appreciated 'that the former might actually be converted into an anti-liberal force.'[1]

It was with this end in view that Bismarck took Prussia to war on three separate occasions. In each case his ulterior purpose was to identify the Prussian state with the cause of German nationalism, thereby appropriating a key aspect of the liberal project and bolstering popular support for the monarchy. In this he was to prove conspicuously successful. Victory over Denmark in 1864 brought the duchies of Schleswig and Holstein into the German orbit, to popular acclaim. Two years later Prussia defeated Austria at the battle of Sadowa, thereby ensuring that the process of German unification would henceforth be centred on Berlin rather than Vienna. And then in 1870 Bismarck succeeded in goading France to declare war on Prussia over the contentious matter of the Spanish succession. Patriotic fervour rallied various minor German states to the Prussian cause, resulting in a wider Franco-German war that Bismarck exploited to elevate (a somewhat reluctant) Wilhelm I of Prussia to the status of German Emperor. Moreover, while doing all this, he successfully avoided precipitating a wider European conflict. For the fact of the matter was that Prussia's neighbours viewed the prime minister's efforts to consolidate German power with intense suspicion. We 'do not live alone in Europe, but with three other powers which hate and envy us' Bismarck observed to his wife in 1866.[2] The possibility of military intervention designed to frustrate his goals therefore had to be taken very seriously indeed. The requirement to remove obstacles to empire without creating new ones in the process: these were the twin political considerations that informed Prussian strategy during the wars of unification.

The armed forces

During the years following the defeat of Napoleonic France, the size and composition of European armed forces returned to something resembling the eighteenth-century pattern. Armies were kept relatively small, the rank and file being either conscripted or recruited for long terms of service, while the officer corps was drawn from among the aristocracy. As such these old-pattern armies reflected the priorities of conservative regimes that felt rather more threatened by popular unrest at home than by aggression from neighbouring states. Keeping armies small meant that the soldiery could be drawn from less politicized sections of the population, and could thereafter be more readily isolated from liberal influences. In this respect the rural peasantry were typically regarded as the most satisfactory source of soldiers, whereas the rapidly expanding urban poor were viewed with deep suspicion, the squalid living conditions of early nineteenth-century cities providing more fertile ground than did the fields for radical politics.

Such was the pattern until the revolutionary disturbances of mid-century abated. Thereafter the closer identification of nation and state worked to change threat perceptions; for although nationalist sentiment helped to defuse tensions at home, it did so only by redirecting discontent outwards against other nation-states. Thus not only did larger armies become politically feasible, in the sense that the masses accepted increasingly inclusive forms of conscription without showing any compunction to turn their military training against the state, but such armies also became more desirable as a hedge against external threats. It was in this context that Prussia could count on rapidly mobilizing half a million soldiers at the outbreak of war with France in 1870.[3] Thereafter, quantitative competition against the backdrop of increasingly tense international relations produced ever-larger armies, with the result that by 1914 around 1.5 million German soldiers could be made almost immediately available in the event of war.

This competitive increase in the size of continental armies was complemented by a parallel form of qualitative competition as the pace of technical change quickened during the nineteenth century. The application of steam power, in the form of the railway, greatly facilitated the growth of armies. Large numbers of troops, along with their weapons and equipment, could now be concentrated at the enemy's frontiers with great speed; and once there they could be reliably supplied so long as they did not move too far from their railheads. The weapons with which they fought were also replaced with improved versions far more frequently than had previously been the case. For not only were superior weapons now consciously sought by private inventors and state arsenals alike, but it also proved possible to field them in militarily significant quantities thanks to new mass-production techniques that achieved productivity rates far exceeding those possible with established artisanal methods. Whereas Prussia took some 26 years (1840–66) to complete its introduction of the *Dreyse* breech-loading rifle, subsequent French efforts to re-equip its army with the *Chassepôt* breech-loader began in 1866 and were completed in just four years.[4]

Prussia's experience explains why the smooth-bore, muzzle-loaded musket, with an effective range of around 100 metres against formed bodies of troops, and a rate of fire of two rounds per minute, had remained the standard infantry arm for the previous century and a half. During the next 50 years it was superseded by a series of increasingly lethal weapons that rolled off the new production lines, culminating in the magazine-fed, breech-loading rifle that boasted an effective range in excess of 1,000 metres and a rate of fire of around 20 rounds per minute. The infantry's firepower was further boosted by the introduction of the machine gun during the late-nineteenth century. The Maxim gun, which was invented in 1885, could spit out bullets at a rate of 600 per minute and was both smaller and lighter than its hand-cranked predecessors such as the US Gatling and the French *Mitrailleuse*. A parallel process replaced the smooth-bore cannon of the Napoleonic era with rifled breech-loaders over the same period, while the introduction of more effective methods of recuperation greatly increased rates of fire. The French 75-mm quick-firing field gun, which was introduced into service in 1897, set the standard for the day with a range of around 7,000 metres and a rate of fire of six rounds per minute. All this meant, of course, that warfare became an increasingly lethal activity as the century progressed – one that threatened to devour the new conscripted masses just as rapidly as they could be fed into battle by the train-load; for battle was indeed placed firmly at the centre of Prussian strategy.

Strategy for unification

The technical considerations outlined above fed into a Prussian approach to strategy that was profoundly influenced by the example of Napoleon I. The Emperor's military genius – his facility for destroying the enemy's armed forces by means of battle – was held up as the model to emulate, and his exploits were subjected to intense scrutiny. There was more to this than simple hero worship: in Prussia the memory of 1806 provided a chilling reminder of the dangers associated with exercising moderation in the conduct of war. Acting in accordance with the strategic orthodoxies of the previous century, the army had been destroyed at Jena–Auerstädt, a victim of Napoleon's superior willingness to risk everything on the outcome of battle at the first opportunity. The result had inspired Clausewitz to conclude that 'he who uses force unsparingly, without reference to the bloodshed involved, must obtain a superiority if his adversary uses less vigour in its application.'[5] This was a sentiment that found receptive ears among his fellow countrymen. On the other hand, Clausewitz's subordination of this logic to wider political considerations encountered a selective deafness among Prussian generals who were concerned to ensure that they would never again be defeated for want of effort. The result was a preference for decoupling the selection of strategic objectives from the political context that gave rise to war.

This preference found practical expression in the generalship of Helmuth von Moltke who, in his capacity as chief of the Prussio-German general staff, provided the military victories that Bismarck exploited to stitch together imperial Germany. While Moltke (citing Clausewitz) acknowledged that war proceeds from political differences, he also strongly espoused the view that having precipitated war, politics should play no role in its conduct per se. Thus:

> Policy uses war for the attainment of its goals; it works decisively at the beginning and end of the war, so that indeed policy reserves for itself the right to increase its demands or to be satisfied with a lesser success. In this uncertainty, strategy must always direct its endeavors toward the highest aim attainable with available means. Strategy thus works best for the goals of policy, but in its actions is fully independent of policy.

In practice Moltke's pronouncement meant that, means permitting, the strategic objective should always be the destruction of the enemy's armed forces.[6] This, it need hardly be observed, was to be achieved via resort to battle.

Two principal obstacles to success in this regard confronted generals during the mid-nineteenth century. One of these was the increasing lethality of rifles and artillery, which meant that lack of care in the commitment of forces to battle would be severely punished – particularly if a frontal attack was the result. With this problem in mind, Moltke considered that the best prospects for success stemmed from carefully coordinated operations calculated to pin the enemy along his front while enveloping one of his relatively unprotected flanks. Under such conditions, he observed, 'Strategy has thus accomplished the best that is to be attained, and great results must be the result.'[7] For Moltke, therefore, strategy was a purely military enterprise focused on manoeuvring one's own armed forces against those of the enemy, in a manner calculated to precipitate battle under the most advantageous conditions.

Indeed in 1866 Moltke kept Prussian forces dispersed until the last moment in order to maximize their chances of enveloping and destroying the more concentrated Austrian army without incurring undue casualties in the process. In the event, his attempted envelopment did not work perfectly – an outcome that serves to highlight the challenges associated with coordinating the operations of large and dispersed forces. This much Moltke himself admitted when, two years later, he declared that 'the plan of offensive operations against France . . . consists simply in making for the enemy's principal force and attacking it wherever we find it. The difficulty lies in carrying out this plan with very large masses.'[8] This second obstacle to success had become serious indeed by mid-century, by which time Prussia could rapidly mobilize rather more soldiers than Napoleon had struggled to employ efficiently in 1812. The friction attendant on their operations was therefore very considerable, and constantly threatened to undermine the carefully coordinated manoeuvres that Moltke viewed as the antidote to modern firepower. In his efforts to manage the problems associated with manoeuvring large, cumbersome and highly vulnerable forces, Moltke was greatly aided by the activities of his general staff, an organization that grew under his auspices into an important ingredient of military victory. It was the general staff that conducted painstaking planning and preparation for the initial stages of Moltke's campaigns, exploiting the telegraph and the railway to achieve the rapid mobilization and concentration of troops that caught their less efficient opponents on the hop. As such it was an important ingredient of his victories.

Nevertheless, Moltke fully appreciated that no amount of planning and preparation could hope to banish friction from the conduct of operations. The effects of chance, uncertainty and a host of other factors would inevitably make themselves felt, over-turning assumptions and undermining certainties. Ultimately, therefore, successful generalship would always demand the moral and intellectual capacity to cope with such twists of fate and to extract benefit from them wherever possible. Indeed, Moltke owed a great deal of his own success to his native military genius. Much like Napoleon, he was readily able to grasp the essentials of a complex and fluid operational situation before acting rapidly and decisively. It was with the authority of personal experience that he argued for success being dependent on 'penetrating the uncertainty of veiled situations to evaluate the facts, to clarify the unknown, to make decisions rapidly, and to carry them out with strength and constancy.'[9]

It was Moltke's capacity for performing such feats that enabled him to commit his forces more handily than the comparatively sluggish Austrians were able to react, thereby precipitating battle under favourable – if not quite ideal – circumstances. Likewise, the superiority of his technique was clearly demonstrated during the opening month of war with France, during which the great frontier battles were fought. As in 1866 Prussian performance was by no means perfect, but what mattered was relative efficiency and in this regard Moltke possessed a vital edge. In the context of what were frequently unanticipated and confusing circumstances he brought the French army to its knees, bottling part of it up in the Fortress of Metz and, after some fierce fighting, forcing the rest to surrender at Sedan – where he also bagged the French emperor, Napoleon III, for good measure.

And yet competent as he undoubtedly was at battling with the influence of friction and the French alike, Moltke encountered a far more intractable problem in the form of Bismarck and his determination to subordinate military operations to political considerations. To 'my shame I have to confess that I have never read Clausewitz and have known little more about him than that he was a meritorious general' the prime minister once remarked.[10] But whatever lacunae existed in his reading, Bismarck was certainly convinced that military operations must be conducted with due regard to the ultimate political goals they were intended to achieve. The 'government of a state engaged in war' he argued

> must look in more directions than towards the scene of the struggle only. The task of the commanders of the army is to annihilate the hostile forces; the object of war is to conquer peace under conditions which are conformable to the policy pursued by the state. To fix and limit the objects to be attained by the war, and to advise the monarch in respect of them, is and remains during the war just as before it a political function, and the manner in which these questions are solved cannot be without influence on the method of conducting the war. The ways and means of the latter will always depend upon whether the result finally obtained is the one desired . . .[11]

More specifically, the army's concern with disarming the opposition must be subordinated to the countervailing demands of peace. To permit otherwise would be to undermine war's utility as a tool of politics. Thus 'the question of war or peace always belongs, even in war, to the responsible political minister, and cannot be decided by the technical military leaders.'[12]

In the event, therefore, Prussian strategy during the wars of German unification was an emergent property of the interaction between military and political imperatives as embodied in the personages of Moltke and Bismarck. For neither of these two men was the experience of making strategy together an easy one. Prussian constitutional arrangements granted equal status to the Chief of the General Staff and the prime minister, a situation that would have benefited from close cooperation between them in the event of war. Cooperation between two such powerful personalities – each supremely confident in his own professional sphere – was never likely to come easily, however. Indeed the outcome of these arrangements was more likely to be an impasse that could only be broken by the intervention of the king. This indeed is exactly what happened in relation to the termination of hostilities with Austria and France, a key question in both instances being whether fighting should cease only once the enemy had been comprehensively disarmed, or whether political considerations warranted a negotiated peace. In this regard it was fortunate for the prime minister that, although Wilhelm greatly admired Moltke, he was ruled by his head rather than his heart and ultimately remained amenable to Bismarck's counsel during both wars.

In 1866 the Prussian victory at Sadowa left the Austrian army retreating on Vienna and temporarily incapable of offering further resistance. The battle also resulted in heavy Prussian casualties, however, without producing their adversary's complete envelopment and destruction. In Moltke's book, therefore, the situation now demanded a prompt continuation of offensive operations designed to finish off the Austrian army. Any prevarication would provide the enemy with a breathing space in which to reorganize and fight back. Bismarck, on the other hand, considered that events at Sadowa had furnished him with sufficient leverage to terminate the war on terms that would preclude Vienna from the project of German unification. This was everything he had wished to achieve. The prime minister was therefore concerned that nothing should be done to stir up unnecessary bitterness between Austria and Prussia, and to this end he opposed any continuation of operations that would lead to a worsening of Vienna's plight.[13] Matters were made more delicate still by Napoleon III who, in the wake of Sadowa, offered his services as an intermediary through whom armistice negotiations could take place. To refuse such an offer would have been to snub a touchy Napoleon, who expected to have a voice in the settlement of European affairs, thereby risking a widening of the war to include France. Under these circumstances,

Bismarck worked hard to restrain the army from undertaking further operations of a kind that would jeopardize peace negotiations with Austria and antagonize the other European powers. Ultimately, however, the king reluctantly acquiesced in his prime minister's wishes, political considerations triumphed over military concerns, and Bismarck had his way. Offensive operations were not renewed and negotiations proceeded in the absence of aggravating military developments.

Bismarck won out over Moltke in 1866 only at the cost of souring relations with the army. Indeed when war broke out with France four years later, the generals demonstrated a remarked reluctance to share information on military-operational matters with him. This became a particular problem in the wake of Sedan, when the question of how best to impose peace once again exposed the difficulties associated with reconciling military and political imperatives. Matters were complicated by the fact that Napoleon's capture had precipitated yet another republican revolution in Paris, and that the new government was determined to continue hostilities with the aid of fresh forces raised in the French interior. Indeed it was in this context that relations between Moltke and Bismarck reached breaking point.

True to form, Moltke favoured a continuation of the war with a view to destroying the newly raised French forces before they could become a serious menace. Peace, he believed, could be countenanced only once the French spirit of resistance had been well and truly crushed wherever it raised its head. Bismarck, on the other hand, feared that a continuation of hostilities along such bloody and protracted lines would ultimately fracture the very coalition of German states that he was hoping to forge into an empire, or that European neutrals would be induced to intervene on behalf of France. If such catastrophes were to be avoided, then the war had to be brought to a more rapid conclusion than Moltke's military prescription permitted. To this end Bismarck considered that an offer of peace on relatively magnanimous terms, coupled with a punitive bombardment of Paris, would provide the correct combination of carrot and stick to terminate the war quickly.[14] For their part, the military demurred on technical grounds. Any such bombardment, they argued, would be a demanding distraction at a time when new threats were emerging in the provinces. But Bismarck stuck to his guns (so to speak). Mindful of the political dangers attending a long and bloody continuation of hostilities, he continued to press for his bombardment. The king once again sided with him, and the prime minister had his wish. Paris was bombarded and surrendered shortly thereafter in January 1871.

Strategy for empire

The emergence of a new German empire from the midst of a victorious war with France sparked renewed worries about the influence of militant nationalism at the heart of Europe. In 1871 Benjamin Disraeli was moved to characterize the Franco-Prussian War as 'the German Revolution, a greater political event than the French Revolution of the last century. . . . '[15] Such alarm was, however, misplaced, for the fact of the matter was that the new empire was a profoundly conservative project with little sympathy for populist sentiment. Bellicose rhetoric aside, Bismarck entertained no Napoleonic flights of fancy. He had saved the Prussian monarchy by creating an empire, and that was entirely enough for him. Thereafter, in his new capacity of German chancellor, he pursued a cautious foreign policy geared towards preserving the new status quo without resort to war. Nevertheless, such a policy could by no means guarantee that Germany would remain unmolested. Moltke therefore planned accordingly, and in the process became increasingly alarmed at the prospect of what another war would bring.

The proximity of so many strong and, in the case of France, vengeful, nation-states oppressed Moltke whose technical prescriptions were now slowly but surely becoming victims of their own success. In 1880 he observed that 'our army stands behind those of our neighbours in point of numbers. It can only make the deficiency good . . . by thorough efficiency.'[16] And yet in the wake of 1871 the advantages stemming from careful military planning and preparation under the direction of a technically proficient general staff were not lost on the other European powers, which thereafter took steps to emulate the German model. As one knowledgeable British commentator subsequently observed: 'That the Prussian system should be imitated, and its army deprived of its monopoly of high efficiency, was naturally inevitable. Every European state has to-day its staff college, its intelligence department, its schools of instruction, and its course of field manœuvres and field firing.'[17] In consequence it became increasingly unlikely that any future war would see Germany achieving the rapid destruction of its adversary's armed forces. On the contrary, matters would be resolved only after a prolonged and mutually exhausting struggle of the kind that Moltke outlined in 1890 during his final address to the *Reichstag*.

> Gentlemen, if war, which now for more than ten years has been hanging like a sword of Damocles over our heads – if war breaks out, one cannot foresee how long it will last or how it will end. It is the Great Powers of Europe which, armed as they never were before, are now entering the arena against each other. There is not one of these that can be so completely overcome in one, or even in two campaigns that it will be forced to declare itself vanquished or to conclude an onerous peace; not one that will be unable to rise again, even if only after a year, to renew the struggle. Gentlemen, it may be a Seven Years' War, it may be a Thirty Years' War; and woe be to him who sets Europe in flames, who first casts the match into the powder-barrel.[18]

Of particular concern to Moltke was the prospect of an alliance between France and Russia, which might well condemn Germany to fight a potentially disastrous two-front war. Lacking the wherewithal to visit rapid destruction on either the French or Russian armies, Moltke planned to divide his strength so as to leave neither adversary entirely unopposed. But such a dilution of effort meant that he could entertain no hope of completely disarming either enemy. In the absence of the necessary means, the 'highest objective of strategy' would therefore have to be something rather more modest. Indeed, in the event of such a war Moltke planned to conduct operations designed to inflict such damage on the enemy as to make a negotiated peace seem preferable to further hostilities.[19] In this context at least, Moltke now found himself agreeing with Bismarck that the strategic objective of his forces could not be 'completely independent' from the political context it which it was formulated. The war, in other words, would need to be terminated through a political process backed by the use of force, rather than by force alone; and to this end Moltke latterly ensured that Bismarck was kept apprised of developments in his planning.

Unfortunately for Germany, this newly harmonious relationship between the technical and political dimensions of strategy was to prove short-lived. Certainly it did not survive the accession of a whimsical and unbiddable Wilhelm II to the imperial throne. The new emperor was intent on introducing a much greater measure of personal rule than had hitherto been usual. Indeed, as one contemporary observer summed up the situation, his 'chancellors have been vice-chancellors; his secretaries of state for foreign affairs have been under-secretaries.'[20] Not surprisingly, Bismarck hardly fitted the mould in this regard and was consequently an early casualty of these new arrangements. This might not have mattered unduly had Wilhelm proved to possess sound judgement of his own, but the facts of the matter were very much otherwise. Wilhelm's bumbling efforts to foster personal relationships with the other crowned heads of Europe proved disastrous, and when spurned he readily sought solace in his self-created role as chief warrior of a beleaguered German nation, the military aspects of his formative years being those for which he retained the keenest sympathy. In the event, Wilhelm's propensity for conflating the demands of his own fragile ego with matters of imperial security would produce dire consequences all round. Having been politically (which is to say personally) rebuffed by his royal relations, he set about attempting to redress matters on the military front. In the process he reduced German strategy to the status of a military-technical exercise in the efficient application of force. Matters would take their most disastrous turn in relation to Russia, which rejected Wilhelm's overtures in preference for an alliance with Germany's implacable adversary France.[21] As a consequence the emperor ordered that new plans be drawn up for exactly that two-front war that Moltke had most feared.

Responsibility for planning such a war fell to Count Alfred von Schlieffen in his capacity as chief of the German general staff between 1891 and 1905. Confronted by the prospect of fighting French and Russian armies that his own forces were not strong enough to tackle simultaneously, Schlieffen sought to identify a way of dealing with them successively and decisively. To this end he decided that the bulk of the German army should be committed against France in the first instance. The French army was capable of mobilizing rather more quickly than its Russian ally, and thus represented the really pressing threat. The trick, then, would be to destroy the French army in rapid fashion (some six weeks were allocated to this goal) before shifting German forces eastwards to deal with the emerging Russian threat. In this regard Schlieffen departed from Moltke who had considered that such a war could not be terminated by military means alone, that political negotiations would need to substitute for the enemy's disarmament. Here once again we see the influence of Wilhelm at work, for although the technical problems faced by Moltke had in many respects grown rather more challenging by the turn of the century, Schlieffen enjoyed a far more permissive political environment in which to develop radical solutions than had his predecessor.

On the technical front, the lethality of the weapons fielded by modern armies meant that German forces could now be expected to suffer truly massive casualties unless committed to battle under the most auspicious circumstances. Moreover, it would avail Germany little if, in defeating France, her own forces were depleted to such an extent that she could no longer manage the

Russian army. It was with this problem in mind that Schlieffen placed the idea of enveloping the French left wing at the centre of his plan – an outcome that was to be achieved by sweeping the bulk of his forces in a great arc through Belgium and Holland. In effect his offensive was conceived as a single, huge version of the envelopment battles that Moltke had sought to engineer in 1866 and 1870. Indeed, like Moltke he believed that an 'encirclement battle' intended to destroy completely the enemy army represented 'the highest achievement of strategy'.[22]

The controversial point here was that both Belgium and Holland would, if left to their own devices, have remained neutral in the event of a war between Germany, France and Russia. By violating their neutrality Germany would therefore risk adding them to its list of enemies. Of course, they might logically have been expected to cede passage to German forces rather than throw away their own tiny armies in a fruitless struggle. On the other hand, violating the sovereignty of a neutral state was illegal under international law and therefore risked precipitating foreign intervention. Belgium's neutral status had, after all, been guaranteed by international treaty since 1839. Moreover, Britain could also be expected to react badly to the prospect of Belgian ports falling into German hands. Certainly a politically astute figure such as Bismarck would have baulked at the prospect of Germany reneging on her treaty obligations and permitting itself to be painted as an unprincipled aggressor. Yet when Schlieffen briefed Wilhelm and his political advisors on his proposed scheme of operations he elicited no objections. On the contrary, the Emperor evidently considered it to be entirely in accord with his own robust preferences for dealing with those who rejected his good offices. This much is strikingly revealed by his Chancellor, Berhard von Bülow, whose memoirs report an astonishing exchange from 1904 between the Emperor and Leopold II of Belgium, to whom Wilhelm declared

> I could not be played with. Whoever, in the case of a European war, was not for me, was against me. As a soldier I belonged to

the school of Frederick the Great, the school of Napoleon the First. As, for the one, the Seven Years' War had begun with the invasion of Saxony, and as the other had always with lightning speed forestalled his enemies, so should I, in the event of Belgium's not being on my side, be actuated by strategical considerations only.[23]

In the wake of this outburst a stunned Leopold departed the scene having 'put on his Prussian Dragoon's helmet back to front!' Bülow then attempted to offer Wilhelm some sage advice along the lines that 'Wars were not won in the long run by military measures alone, but also by political considerations. Napoleon ended as a prisoner in spite of his military genius', only to be told that such thoughts would disqualify him from the post of Chancellor in the event that Germany went to war.[24]

In such a climate it is hardly surprising that Schlieffen was able to focus exclusively on the technical dimension of strategy, and indeed on solving the challenges associated with fighting an envelopment battle of such heroic proportions. For whereas Moltke's battles had each involved a few hundred thousand soldiers, Schlieffen was thinking in terms of a million and a half, and herein lay another serious problem. The task of directing such large forces in a lightning offensive, designed to envelop and destroy the entire French army, demanded a degree of choreography that even Napoleon I or Moltke would have struggled to impose. The sheer scale of the operation would provide great scope for inefficiencies to creep in, leading to a loss of all-important speed. Schlieffen was of course acutely aware of this problem, and sought a partial answer in the new techniques of command and control that were emerging at the time. Under modern conditions, he wrote:

No Napoleon stands upon a rise surrounded by his brilliant retinue. His white horse would be the easy target of countless batteries. The *Feldherr* [commander in chief] finds himself further back in a house with a spacious office, where telegraphs, telephones and signals apparatus are to hand and from where fleets of cars and motorcycles, equipped for the longest journey, patiently await orders. There, in a comfortable chair before a wide table, the modern Alexander has before him the entire battlefield on a map. From there, he telephones stirring words [*zündende Worte*]. There, he receives reports from the army and the corps commanders, from the observation balloons and from the dirigibles that observe the movement of the enemy along the whole line and that look behind the enemy's positions.[25]

This, however, was a vision of warfare that turn-of-the-century technique was incapable of realizing. In practice, efforts to exert effective command and control over German forces would be attended by difficulties that continued to resist solution for decades to come. In consequence, the success of Schlieffen's gigantic operation always relied on painstaking planning and preparation by the German general staff in an effort to anticipate and eliminate as many 'frictional' barriers to the efficient conduct of the offensive as possible. Whereas Moltke had famously warned that no plan can be expected to survive in the face of enemy resistance, Schlieffen was relying on operations proceeding so smoothly and effectively that the French would be denied the time necessary to react and disrupt proceedings.[26]

The 1914 offensive

When Germany finally went to war she did so with Moltke's nephew as chief of the general staff. This younger Moltke had earlier entertained reservations about the rigidity of Schlieffen's original conception, which he considered rather too audacious and dangerously inflexible should matters not unfold entirely as hoped for. Accordingly he introduced some significant changes, reining in his right wing so that it would no longer pass through Holland, and strengthening his centre against which he feared the

French might mount an offensive of their own. In this regard he faithfully echoed his eponymous uncle's famous distrust of rigid plans, although it would be to Germany's misfortune that he was not similarly capable of capitalizing on the fleeting opportunities offered by the fluid operational situation that presented itself in 1914.

Once the German right wing entered Belgium it encountered unexpected resistance from its tiny army, combined with disruptive acts of sabotage by disgruntled Belgian civilians. While none of this was sufficient to halt the German advance it did impose delays of a kind that Moltke could ill-afford. Moreover, as his forces swept in their great arc through Belgium and into France, the baleful influence of friction made itself increasingly apparent. Day after day of forced marching left soldiers exhausted and ill-provisioned; reliable information about the status, location and intention of enemy forces proved difficult to obtain, leading to dangerously optimistic assumptions supplanting sober facts; and tactical initiatives taken by formation commanders further reduced Moltke's capacity for exerting control over the situation. Thus as German forces pushed doggedly on towards what they believed to be imminent victory, they were actually exposing their right flank to a major French counter-offensive on the Marne. Moltke belatedly realized the danger, but his corrective orders (issued via the newfangled wireless that had so captured Schlieffen's imagination) arrived too late to rectify matters. The French counter-offensive duly went home, sending a startled adversary reeling. From this point on, there was no hope of Germany winning a rapid victory in France.

According to Gerhard Ritter the 1914 offensive never enjoyed any real prospect of success.

The great Schlieffen Plan [he observed] was never a sound formula for victory. It was a daring, indeed an over-daring, gamble whose success depended on many lucky accidents. A formula for victory needs a surplus of reasonable chances of success if it is to inspire confidence – a surplus which tends quickly to be used up by 'frictions' in the day-to-day conduct of war. The Schlieffen Plan showed an obvious deficit in these chances . . . [27]

This is overstating the matter somewhat. To be sure, by 1914 the French army was no longer the hopelessly disorganized effort of 1870 relieved only by the bravery of the soldiery and the lethality of their new rifles, but a highly efficient force capable of rapid mobilization and commitment to action. Indeed, when war broke out almost two million soldiers were rushed into place by 4,278 trains, of which just 19 were delayed.[28] In the face of such efficient action, a German victory was always going to be difficult to achieve. Nevertheless, Moltke's revised arrangements might *just* have worked in the hands of a more capable wartime general – such as his uncle had proved himself to be. Even at a late stage in the offensive there remained time to shift forces away from obviously unproductive operations in the German centre and onto the crucial right flank, where their presence might have proved decisive. Had the younger Moltke possessed his uncle's capacity for readily grasping the essentials of a rapidly unfolding operational situation, and for responding decisively to the opportunities presented, he might have led Germany to a startling victory in 1914. But while Moltke certainly appreciated the virtue of flexible plans, he proved indecisive in time of crisis and thus incapable of capitalizing on the flexibility he had succeeded in providing for himself.

A more fundamental criticism of pre-war German strategy was the extent to which it discounted the wider political context in which it was made. That the stubborn character of Belgian resistance came as a surprise is indicative of the extent to which technical considerations dominated thinking. Belgium ought not to have fought because the correlation of forces was so obviously against it – and yet it did. More importantly, the turning movement through neutral Belgium effectively paved the way for British military intervention. In the absence of such a casus belli the question of how to respond would certainly have split Britain's ruling liberal party and led to a paralysing political crisis. In the event Germany's actions rendered the point moot, and an expeditionary force of six divisions was despatched to the Continent. Of course, six divisions amounted to little more than a 'speed bump' in the path of the German juggernaut. On the other hand their commitment made it more likely that, the duration of hostilities permitting, others would follow in their wake. And this indeed is exactly what happened: over time the six divisions of the British Expeditionary Force grew by a factor of ten, in the process greatly multiplying German problems on the Western Front. In such a manner did Germany's emphasis on military-technical imperatives, at the expense of wider political considerations, act to shift matters even further to her own disadvantage.

Had the slender German forces masking Russia's mobilization not performed unexpectedly well against their more ponderous adversary, Germany would have soon found itself occupying a difficult position between a rock and a hard place. As it was, an unexpected victory at Tannenberg kept the Russians at bay, while operations in the west took on a new form. With the initial fluid phase of attack and counter-attack over, the German and Anglo-French armies extended themselves in the search for an enemy flank to envelop. In the event, however, the forces on both sides proved so large as to permit the creation of a continuous front stretching from the Channel coast to the Swiss border. Under such conditions, traditional envelopment operations proved impossible, and the armies were forced to confront each other head on. The resulting battles were bloody in the extreme as the new artillery and machine guns whittled away at both sides, converting the war into a huge attritional struggle. Against this background technical developments – including aircraft, tanks, gas and new doctrines for attack and defence – contributed to the increasing efficiency of operations. By the war's end, a more extensive use of radio made it possible to harness these innovations to a modern conception of combined-arms operations that constituted a remarkable departure from the pattern of 1914. Nevertheless, such developments did not generally remain the monopoly of one side or the other for very long. Rather, they were more or less simultaneously achieved in the context of an ongoing struggle for technical supremacy, with the result that they cancelled one another out. Ultimately, therefore, victory turned on the capacity of each belligerent to make good the huge casualties associated with battle; and ultimately it was Germany that ran out of soldiers first, a victim of the numerically superior coalition that two decades of disdain for the political dimension of strategy had helped to assemble against herself. In the process rather more than 15 million casualties were suffered on all sides.[29]

The highest form of strategy

Looking back over the events that gave rise to this calamity, Hajo Holborn observed that the 'highest form of strategy is the outcome of military excellence enlightened by critical and instructive political judgement.'[30] The more specific point he was concerned to make was that, although Prussio-German strategy was generally technically excellent, it was less often politically enlightened; and it would be difficult to disagree with him in this regard. A high point was reached under Bismarck who, according to Isaiah Berlin, possessed an unusually sure grasp of the 'public medium in which he acted', which is to say 'the potential reactions of relevant bodies of Germans or Frenchmen or Italians or Russians'.[31] Indeed, it was his judgement on such matters that made war into a practicable means of pursuing German unification in the teeth of European suspicion and rivalry; for Moltke's technical focus on the application of force would, if left to its own devices, certainly have created more political problems than it solved. Moreover, Bismarck was wise enough to perceive that the empire he had created in 1871 must thereafter seek to defend itself via a policy that sought to avoid war rather than impose its will via military means. Michael Howard is surely correct in his view that it 'was thanks entirely to Bismarck's statesmanship that Moltke's victories were not to remain as sterile as Napoleon's, but were to lead, as military victories must if they are to be anything more than spectacular butcheries, to a more lasting peace.'[32] Thereafter matters went rapidly downhill. The technical content of Wilhelmine strategy remained as strong as ever, but this proved to be a wasting asset in the wake of 1871 as the other European powers worked to close the gap in this regard. Under such conditions, no amount of technical preparation would suffice to make up for the fact that, unlike Bismarck, the new emperor possessed no firm grasp of the 'public medium in which he acted'. Germany badly needed the allies that it lost as a result of his antics, just as it badly needed a strategy that would not further add to its list of enemies in the event of war. The collapse of the political dimension of Wilhelmine strategy was therefore a disastrous development indeed.

3 Total war and liberal dissent

In 1898 the Polish financier Ivan Bloch published his massive six-volume work, *La Guerre Future*, parts of which were translated into English under the title *Is War Now Impossible?* Bloch's central claim was that technical developments had now made war so destructive as to render it a suicidal undertaking.

> The very development that has taken place in the mechanism of war has rendered war an impracticable operation. The dimensions of modern armaments and the organisation of society have rendered its prosecution an economic impossibility, and, finally, if any attempt were made to demonstrate the inaccuracy of my assertions by putting the matter to a test on a great scale, we should find the inevitable result in a catastrophe which would destroy all existing political organisations.

More specifically, he anticipated that recent increases in the lethality of small arms and artillery would result in the creation of deep zones of fire-swept ground that troops would find impossible to traverse without incurring tremendous casualties in the process. Armies would therefore find it correspondingly difficult to close with one another, and any chance of achieving a swift victory would vanish as a result. Instead, military operations would degenerate into a prolonged bout of mutual attrition, dominated by the effects of firepower and the consequent need for extensive entrenchments, with the 'dying and the dead . . . utilised as ramparts to strengthen the shelter trenches.'[1]

And if battlefield conditions could be expected to be unpleasant in the extreme, the broader consequences of prolonged war were even more alarming. Bloch predicted that the belligerent nations would simply be unable to afford the crippling costs associated with protracted hostilities, while the great mass of conscripts that had been called up to fill the ranks could not be spared indefinitely from their farms, as the civilian population would still need to be fed. This latter problem would, moreover, be compounded by the deleterious effects of war on established patterns of trade because no European nation (with the exception of Russia) was self-sufficient in terms of food. Starvation was thus seen by Bloch as an inevitable consequence of war, with the dread prospect of revolution lurking menacingly in its wake.

Although Bloch accurately predicted the strategic problems flowing from the military-technical developments of his day, he was wide of the mark in relation to the economic and political consequences. Despite the lengthy character of the First World War there was no generalized European collapse of the kind he had feared. To be sure, national wills and economies were severely strained by the remorseless consumption of manpower and materiel, but there was no dramatic collapse behind the front lines until that of imperial Russia in 1917. One reason for this was the success with which governments intervened in the running of their wartime economies. Rationalization of war production, combined with the strict regulation of civilian consumption, eked out resources to the extent that the end of hostilities was consequent on military victory rather than on mutual exhaustion. Whatever its generals may subsequently have claimed, the German army had been stretched to the limit by late 1918 and would certainly have collapsed under the pressure of another Allied offensive had not the Armistice intervened.

Total war
That said, the fact that thorough-going mobilization had proved capable of sustaining the demands created by war did little to lessen the anxieties provoked by the anticipation of future conflict. That the costs of 1914–18 had proved manageable did not mean that they had been borne lightly, or that another war could be

contemplated with any sense of equanimity. Indeed, the new concept of 'total' war that emerged during the interwar period suggested that the costs occasioned by a future European war, and the strains imposed on the belligerents, would be even greater than had previously been the case. This new concept appears to have emerged in France during the First World War, in reaction to the unprecedented degree of economic mobilization required to sustain hostilities. Nevertheless, the person who ultimately did most to popularize it was General Eric Ludendorff, by means of his book, *Der Totale Krieg* (1935).[2]

Ludendorff had initially made a name for himself by dint of leading a successful assault on the fortified Belgian city of Liége during the opening phase of the German offensive in 1914. Thereafter he had been transferred to the East, where he was credited with destroying the Russian army at Tannenberg.[3] In 1916, with the war in the west locked in an increasingly bloody stalemate, Ludendorff was summoned back to Berlin and received by Wilhelm II, who once more exercised a baleful influence over events by effectively giving him carte blanche to direct Germany's war effort in the capacity of First Quartermaster-General. This was a rank of Ludendorff's own choosing and one whose title he evidently took literally, for under him Germany was indeed converted into a great quartermaster's store whose manpower and resources he single-mindedly subordinated to the pursuit of military victory. In the process he ran rough shod over constitutional arrangements, exploiting his position at the centre of the war effort to remove the politically moderate chancellor, Theobald von Bethmann-Hollweg, and to marginalize Wilhelm from the decision-making process.

The centralization of executive power in the hands of a single military figure was necessary, Ludendorff subsequently argued, because the course of the war had revealed that victory hinged on the capacity to sustain an undivided national effort. The new technical conditions under which war was fought had indeed transformed the initial clash of armies into a grim Bloch-style struggle that had sucked in the manpower and resources of the combatant nations, as they each endeavoured to destroy the other's capacity to continue making war. For Ludendorff it followed from this that the fortunes of nations would henceforth rest on the capacity to wage similarly long and highly destructive conflicts. The human and material costs associated with waging such wars would be extreme, but there would be no choice other than to bear them if the alternative was subjection of the kind that Germany had suffered after 1918. The overriding preoccupation of the state should therefore be to ensure that exhaustive preparations were made to gain victory in the next round of hostilities, so that there would be no national collapse under the strain of war. Such efforts could not, moreover, wait until the outbreak of hostilities, as by then it would be too late to cope with an adversary who already had extensive preparations in place. Thus victory would require not only the utmost efforts in wartime, but great sacrifices in peacetime too. Total ends demanded total means.

Such ideas threatened war's political instrumentality in two ways: not only would the costs of fighting be disproportional to any goal short of national survival, but the demands associated with preparing for war would also dominate even the peacetime life of the nation-state. Indeed, for Ludendorff, the traditional relationship between politics and war would henceforth be reversed. 'All the theories of Clausewitz should be thrown over-board' he argued. 'Both warfare and politics are meant to serve the preservation of the people, but warfare is the highest expression of the national "will to live", and politics must, therefore, be subservient to the conduct of war.'[4] In short, Ludendorff revealed himself as the quintessential militarist, in the sense that his highest ideal was a nation trained, equipped and prepared for victory in an existential struggle. To this end he pressed his case for the creation of a 'technical dictatorship for purposes of the conduct of mass warfare.'[5] All else, he considered, could be sacrificed to this goal without compunction.

Liberal dissent

While a 'technical dictatorship' might have been deemed admirable by those with a particularly martial turn of mind, to the liberals of the interwar period such a vision was unpalatable in the extreme. It was, moreover, one that seemed all too likely to be realized as the menace of Nazi Germany grew on the horizon. In the face of such a threat prospects seemed bleak indeed for the European democracies; for not only would efforts to defeat a nation-state whose very ideology placed the idea of struggle at the centre of human affairs be inordinately costly in terms of blood and treasure, but in the process they would demand unprecedented levels of social mobilization. As the British strategist Basil Liddell Hart observed, one likely casualty of all this would be

> the British tradition of individual freedom, our most precious heritage, which will be immediately endangered if we accept the new foreign theory of totalitarian preparation for war. It would be the supreme irony of our history if we sacrificed this freedom in the process of preparing to defend it. It would be like committing suicide to escape fear.[6]

Liddell Hart did not altogether reject war as a political instrument, but his concerns about the threat to individual liberty posed by a totalizing struggle were very much shared by a pacifist movement that was burgeoning in Britain during the 1930s. In a heated exchange with George Orwell (who fervently supported the war against Hitler), the poet D.S. Savage itemized the many and various ways in which fighting Germany was likely to become a self-destructive act. Fascism, he contended, involved the

> curtailment of individual and minority liberties; abolition of private life and private values and substitution of State life and public values (patriotism); external imposition of discipline (militarism); prevalence of mass values and mass-mentality;

falsification of intellectual activity under state pressure . . .
Don't let us be misled by *names* . . . It's the reality under the
name that matters. War demands totalitarian organization of
society. Germany organized herself on that basis prior to
embarking on war. Britain now finds herself compelled to
take the same measures after involvement in war. Germans
call it National Socialism. We call it Democracy. The result is
the same.[7]

Savage was entirely correct to point out the dangers to national
life associated with the demands imposed by total war. But even
so, in the face of a threat such as Nazism, it was difficult to understand how pacifism offered a viable way forward. Adherence to
such a line offered merely to delay defeat by retreating from the
more unpleasant manifestations of international politics until (in
the Nazi case) they finally came hammering on one's door in the
early hours of the morning.[8] In the long run, the result of such an
approach could only be total subjection.

A technical fix?

All this suggested that a major challenge facing the European
democracies was to identify a middle way between the two
extremes of totalitarianism and pacifism, but how might this be
achieved? One possible solution lay in the pursuit of a technical
fix that would remove the requirement to choose between two
equally unpalatable political futures. In this regard the ideas associated with an eclectic group of military theorists, who advocated
future wars being fought by highly mechanized forces, became
the focus of vociferous debate. This group included the likes of
J.F.C. Fuller and Liddell Hart in Britain; Giulio Douhet in Italy;
Charles DeGaulle in France; and Heinz Guderian in Germany.[9]
As a group they held a diverse range of views on the vexed issue
of future warfare, and the value of mechanization therein. For
some the issue was never more than a narrow military-technical
one, in the sense that mechanization was seen as merely the latest

phase of qualitative competition that needed to be addressed if a dangerous disparity in means were to be avoided. Most, however, saw in mechanization the possibility of reducing the costs of future wars to more bearable levels, while some – most notably Liddell Hart – sought to bring such matters to bear on the challenges posed by war for liberal democratic ideals.

In all cases, however, the starting point was a common vision of how an ambitious programme of military-technical innovation would greatly reduce the burden of friction attending the conduct of operations, thereby providing the capacity to inflict extremely rapid defeat on one's adversary. The key to exploiting this capacity involved embracing some ambitious new operational concepts. In the past, it was argued, victory had required the infliction of physical damage on an adversary's armed forces in an attempt to disarm him, thereby rendering him unable to resist the victor's will. Recently, however, scientific developments had made it possible to substitute this slow process of physical destruction with one of rapid psychological dislocation. More specifically, the techniques of mechanization – in the form of tanks and aircraft – now permitted attacks to be mounted directly against the enemy's *will* to fight, thereby avoiding the costly business of destroying his physical means of fighting. Tanks possessed the mobility necessary to avoid the 'teeth' elements of a relatively cumbersome enemy army composed of foot-bound conscripts, and to strike blows against its command and control structures. Attacks of this kind would generate catastrophic levels of friction for the enemy by sowing confusion and promoting moral collapse, thereby depriving him of his capacity to offer organized resistance, without actually killing too many people in the process. More ambitiously still, attacks mounted by aircraft against enemy cities held out the prospect of creating such general panic and confusion as to paralyse a nation's war effort in short order and with nothing like the level of fatalities that the traditional approach to the conduct of war demanded. What mattered, therefore, was to acquire the new weapons in adequate numbers, to organize them into appropriate formations, and to man them with personnel who were well trained in accordance with a set of doctrines focused on delivering rapid victory. This new breed of personnel would themselves need to be technical specialists, which in turn demanded that they be long-service professionals as opposed to short-service conscripts. Like the machines they manned, therefore, they would be expensive to train and maintain. On the other hand they would be relatively few in number. Moreover, liberals saw in them an attractive alternative to training mass conscript armies in peacetime – with all the attendant economic and political costs this would entail – before sending them off to mass slaughter once war broke out.

The limits of technique

While there was certainly merit in these bold visions of future warfare, it was also easy to overestimate the capacity of military technique to generate rapid victories at low cost. None was more optimistic in this regard than Fuller, according to whom

> weapons, if only the right ones can be discovered, form 99 per cent of victory. . . . Strategy, command, leadership, courage, discipline, supply, organisation and all the moral and physical paraphernalia of war are as nothing to a high superiority of weapons – at most they go to form the 1 per cent which makes the whole possible.[10]

To the extent that his fellow travellers shared this enthusiasm, events would prove them to be sadly in error. The basic problem was highlighted by none other than Ludendorff, according to whom technique provided no real solution to the costliness of future warfare. Ludendorff was certainly interested in fielding the most efficient armed forces possible, with a view to disarming the enemy as rapidly as was practicable after the outbreak of hostilities; for the more quickly he was disarmed the less opportunity would he have to inflict damage on friendly forces. To this end he was interested in developments in military technique associated with

mechanization. By the same token, however, he remained unconvinced that technique would itself generate decisive advantages in
a future war. Military-technical competition between states meant
that superior technique would constitute a fleeting advantage that
could not be relied on.

> As in the World War, 'technique was set against technique', so
> men have so far always known how to oppose technical means
> of defence to technical means of attack . . . It may be said that
> the mutual competition of technical means gradually leads to
> an equalization of means of attack and defence or to the
> finding of means for checking them.[11]

It followed from this that sound military technique was a valuable
– even necessary – asset but that, in and of itself, it could not produce victory. On the contrary, any future war's outcome would
ultimately be determined by hard fighting – fighting that demanded both the will and the capacity to inflict and endure great
casualties and material damage over an extended period.

Much the same arguments were employed in Britain by Victor
Wallace Germains, in the context of a carefully considered rejoin-
der to Fuller's early technophilia.

> The theory that success in war is to be gained neither by a
> process of hard fighting, nor by superior leadership, training, or
> numbers, but by the surprise use of some wonderful invention,
> is a very attractive one . . . But it is necessary to remember that,
> if you are pitting yourself against a highly industrialised state,
> the men on the other side are neither fools nor sluggards. They
> will be just as active in research and experiment as you are.

Consequently, technical initiatives taken by one side would galvanize the introduction of countermeasures by potential adversaries,
with the effect of reinstating the original balance of technique.
This led Germains to conclude that the 'theory that the next war

will be a "high-speed" war seems much open to question.'
While technical innovations would not be without value, their
introduction would not substitute for 'the concentrated strength
of the national manhood, and the national resources as a whole,
which . . . can alone bring victory in the next war.'[12]

As such reasoning suggests, it was entirely possible to construct
a logical argument for the case that radical military-technical
innovation would provide no real solution to the challenges posed
by total war. It was reasonable to suppose that tanks would retain
the capacity to deliver rapid and conclusive results only against an
adversary that was unprepared to counter the threat they represented, that otherwise the results would be inconclusive with
operations tending towards deadlock and attrition. Of course it
was always possible to reply that the challenges posed by anti-tank
defences could themselves be overcome by appropriate developments in the technique of tank warfare. Guderian, for example,
argued that tanks committed en masse, over suitable ground and
with the benefit of surprise would be capable of overrunning
enemy defences without suffering crippling losses in the process.[13]
Still, such expedients could not provide a definitive answer to the
challenges posed by anti-tank defences. Rather, they represented
but one small step in an open-ended process of competition.
What came to be called 'blitzkrieg' tactics did indeed augment
the potency of the tank, but it was only a matter of time before
countervailing technical developments robbed them, in their turn,
of much of this potency.[14] In short, the self-cancelling character
of military-technical innovation meant that tanks and their
operations became part of each state's total-war effort rather than
a substitute for it.

Throughout the interwar period it remained rather more
difficult to envisage effective means of blunting air attacks against
cities of a type that might rapidly collapse the national will to
fight. Nevertheless, it was also realized that if both sides possessed
air forces capable of mounting such attacks neither could be considered to possess a significant advantage. Thus unless or until one

side succeeded in destroying the other's air force, thereby gaining what Douhet termed the 'command of the air', such a war might well take the form of a ghastly race to break the adversary's civilian morale via bombing, while the opposing armies grappled with one another in long, costly and largely irrelevant operations of their own. Under such conditions, victory would go to the state whose people were more willing to bear the costs associated with such warfare.

In point of fact it was a similar line of reasoning that eventually led Fuller to question the ability of democratic states to prevail in war against what he perceived to be the superior discipline inculcated by fascism. Initially one of the foremost proponents of exploiting military technique as an alternative to the costs of war, he subsequently performed a veritable volte face by embracing the view that tank warfare would prove indecisive against a similarly equipped adversary, and that the outcome of future wars would henceforth turn on civilian capacity to endure the rigours of aerial bombardment. Indeed by 1937 he was emphasizing

> the question of discipline, because in this present age the spiritual and moral forces are once again rising in the ascendant. . . . In war as it will be waged we want A.1 minds, A.1 hearts, A.1 bodies and A.1 weapons. And though I have said that weapons may represent 99 per cent of victory, without A.1 minds they are little more than scrap iron.[15]

Victory, in other words, would come only to those nations that thoroughly prepared themselves for a long and costly struggle, which would be determined in the moral rather than the technical sphere. To the extent that fascism did indeed excel in the inculcation of discipline, the outcome for liberal democracy looked bleak in an era of total war. To skimp on the necessary preparations would be to lose when war finally broke out, while to make such preparations during peacetime would be to surrender to totalitarianism even before the fight had begun.

Liddell Hart and limited war

By the mid-1930s, therefore, serious doubts were being expressed about the capacity of military technique to facilitate rapid, low-cost victories. The self-cancelling character of military-technical competition made a repetition of Bloch's stalemate appear likely, leading to a prolonged and extremely costly struggle that would demand total mobilization. Under such conditions even a victorious war might well prove to be a self-defeating proposition for democratic states.

A notable exception to this line of thought was Liddell Hart, who persisted in looking at matters rather differently. To be sure, he now dissented from his earlier position on the capacity of mechanized forces to win wars in short order, arguing that recent technical developments militated against such an outcome.

> The trend of modern weapon development has been predominantly in favour of the defence. It was the machine-gun which, above all, established the superiority of the defensive in the last war. To-day there are more machine-guns than ever. The anti-tank and the anti-aircraft gun, the weapons which have been most improved since the war, are purely defensive.[16]

He differed from other commentators, however, by arguing that the Bloch-style stalemate that such a situation could be expected to create might be turned to the advantage of a 'non-aggressive state in war, a state concerned only to maintain its own interests and those of its friends.'[17] This was because the full costs of modern warfare stemmed from thorough-going efforts to disarm one's adversary in the face of competent resistance. States fighting in order to preserve the existing political order might, therefore, avoid paying these costs, and indeed escape the requirement for total mobilization, by waging war in a correspondingly restrained manner. The trick was to identify an appropriate strategic objective; for if it should not be to disarm one's enemy as rapidly as possible, what then could it be? Here Liddell Hart supplied a thoroughly heterodox alternative: the strategic objective should be

> to convince the enemy that he has nothing to gain and much to lose by pursuing a war. Its guiding principle is to eschew the vain pursuit of a decision by the offensive on our part. Its method is not merely to parry, but to make the enemy pay as heavily as possible for, his offensive efforts. This implies in the military sphere an active and mobile defence, in which the effect of direct resistance is extended by ripostes . . . as well as by continual harassing action.

The strategic objective should, in other words, be to preserve one's own forces while wearing down those of the enemy, with a view to convincing him that the costs associated with continuing to fight would outweigh any political gains he hoped to make. Faced with such a situation, an opponent would sooner or later lose heart, for there 'is nothing more demoralizing to troops than to see the corpses of their comrades piled up in front of an unbroken defence, and that impression soon filters back to the people at home'.[18] Peace would therefore follow without the bloodbath associated with an unrestrained effort designed to disarm one's enemy outright.

Liddell Hart was therefore advocating a strategy of limited war, by means of which a democratic state might 'maintain its own interests and those of its friends' in the face of aggression by revisionist powers. Specifically, Britain and France might hope to defend themselves against Hitler's Germany without suffering appalling casualties and sacrificing their own political institutions to the demands of total war. Certainly, geographical and technical conditions reinforced the practicability of his ideas in this regard. Britain's insular position provided her with a relatively secure base from which to prosecute such a limited war. France shared a common border with Germany and was thus far more vulnerable to attack by ground forces. On the other hand the border was one that could be, and indeed was, heavily fortified during the 1930s with a view to rendering any such attack disproportionately expensive. The result was the Maginot Line, of which Liddell Hart talked in positive terms.[19]

The prospects for defending cities against air attack were less certain, and Liddell Hart was on shakier ground when he argued that the anti-aircraft gun had got the better of the aircraft. He was correct, however, to decry the capacity of European air forces to reduce major conurbations to rubble in short order.

> If the mind of Europe is turning ever more to the air, its fears are certainly inflated by a quantity of 'hot air'. So far as the civilian masses are concerned, their present danger is undoubtedly being exaggerated. The reason is simply explained: the air forces of Europe to-day are not large enough to carry out the universal devastation that is popularly imagined.
>
> It seems a fairly safe calculation that the tonnage of high-explosive bombs required to destroy any large city far exceeds the capacity of any country's existing bombers.

Bombing fleets would, of course, eventually become rather more fearsome instruments of war, but defensive techniques against air attack also made impressive gains over the same period, not least because of the introduction of radar in this context. In the meantime Liddell Hart was more impressed by the capacity he discerned in airpower for disrupting the movement and supply of ground forces as they sought to concentrate for battle. This new capacity, he was inclined to believe, represented another reason why major offensive efforts designed to disarm one's adversary were doomed to failure; for the delicate logistical structures on which they depended would not survive in the face of sustained air attack.[20]

All this suggested that Britain and France were in a strong position relative to Germany, and that this position could be exploited via a limited, coercive strategy designed to punish aggression, thereby removing the requirement to prepare for and wage total war. Indeed, Liddell Hart's prescriptions were influential with the Chamberlain government, which was itself anxious to avoid a repetition of the First World War, and which saw in them a means of limiting British involvement in any future European conflict. Why was it, then, that when war finally broke out Hitler succeeded in over-running France and placing Britain in the most desperate of strategic circumstances? In 1937 Liddell Hart had observed that 'if we exclude a policy of non-resistance, any security will depend on our judgement of technical factors as well as political factors.'[21] As far as the military-technical balance was concerned his own judgement was sound enough: under prevailing conditions, efforts to disarm one's enemy in the event of war could reasonably be expected to be extremely costly. On the other hand, the degree to which this state of affairs could be exploited in the formulation of Anglo-French strategy also depended on the political flavour of the enemy. And on the matter of Hitler's motivations, Liddell Hart, just like Chamberlain, was distinctly awry. 'Herr Hitler' was not the reasonable, calculating statesman that both of them evidently believed him to be.[22] Consequently, the prospects for fighting a successful limited war against Germany were far more remote than technical considerations alone suggested. According to Orwell, whose judgement was sounder on such matters:

> 'Limited aims' strategy implies that your enemy is very much the same kind of person as yourself; you want to get the better of him, but it is not necessary for your safety to annihilate him or even to interfere with his internal politics.

The problem with Hitler was that he represented a very *different* kind of person: one who pursued his nightmarish political ambitions with the 'fixed vision of a monomaniac'. Moreover, his hold over the German people was evidently such that they were willing to embrace his offer of 'struggle, danger and death' in an attempt to realize their Fuhrer's vision.[23] The upshot was that war with

Nazi Germany was not going to be resolved via a limited coercive effort, because Nazi Germany was not the kind of nation-state to be intimidated by the prospect of heavy casualties. Ultimately, therefore, the survival of European democracy depended on removing Hitler from power, which as a prerequisite required an unlimited military effort designed to render Germany comprehensively defenceless.

Fighting the war

Over the course of the war that followed, Liddell Hart was proved correct in his appreciation of technical factors, just as Orwell was proved correct in his political analysis. To be sure, in 1940 French resistance rapidly collapsed in the wake of a remarkably daring German offensive. This startling victory was achieved by means of mechanized forces, which mounted a surprise thrust through the heavily wooded Ardennes and across the river Meuse. In doing so, they bypassed the northern flank of the Maginot defences and were subsequently able to plunge rapidly behind Anglo-French lines, causing panic and confusion in their wake. Allied resistance collapsed less as a result of material losses, and principally due to the psychological paralysis induced by the speed of the panzers' advance – much as the more enthusiastic theorists of the 1930s had predicted. In consequence, the Battle of France was seen as validating German reliance on the achievement of rapid victories by means of high quality, but relatively small, mechanized forces.

But the fall of France was not really the vindication of mechanized warfare that that it was subsequently held to be. Having precipitated an Anglo-French declaration of war by invading Poland, Hitler felt it vital to achieve an early victory in the west in order to avoid being overwhelmed by superior numbers and resources in the context of a prolonged struggle. To this end, he cajoled his rather more cautious generals into risking all on a lightning offensive whose prospects for success might reasonably have been considered slim. A large-scale movement of armour

Total War and Liberal Dissent

through the Ardennes, followed by an opposed river crossing was an extremely risky endeavour that might easily have come to grief. The French troops holding the west bank of the Meuse were not particularly strong, but they were powerful enough to crush the bridgeheads that the Germans succeeded in establishing across the river. Indeed, they probably would have succeeded in doing so had they not been plagued by a series of misfortunes that fatally delayed their commitment to battle. In short, the Germans were lucky as well as good, while the French were unlucky as well as being distinctly mediocre. This means that things could very easily have gone differently. Had the panzers been held up for just a little longer at the Meuse, they would probably never have got across. Instead they would have been ground down in a battle of attrition as French reserves were slowly but surely funnelled into the sector, or pounded to pieces by air attack as they struggled to retreat through the Ardennes. In the wake of such events, it is unlikely that daring ground offensives by mechanized forces would have figured highly in anybody's views on how to prosecute the war to a successful conclusion.

Moreover, although Britain had been forced off the Continent in 1940 she remained relatively secure behind the Channel. The strength of the Royal Navy ensured that Germany could not contemplate an amphibious assault without first achieving air superiority over the narrow sea. To this end, the Luftwaffe was tasked with destroying the Royal Air Force (RAF) by attacking its airbases and luring its aircraft into costly battles. Although success in this regard was confidently expected, the Luftwaffe failed to achieve its strategic objective. Whereas the RAF enjoyed no superiority either in the quality or quantity of its aircrews and aircraft per se, Britain could boast the world's first integrated air-defence system. A chain of radar stations, linked to a sophisticated command-and-control apparatus, allowed fighter aircraft to be committed to battle with unprecedented efficiency, thereby permitting them to inflict heavy casualties among the German raiders who were operating under a much greater burden of friction. By such means the Luftwaffe was beaten off and Britain was safeguarded from the threat of imminent invasion.

On the other hand, Britain faced severe difficulties in striking back at Germany from behind the Channel. The aerial bombardment of economic targets represented one possibility in this regard. Now, however, the tables were turned on the RAF, and earlier doubts about the capacity of air forces to achieve meaningful results against such targets were once again borne out. British aircraft proved too few in number, and lacked the range and bomb-load necessary to deal anything like crippling blows against the German war economy, while they also emerged as woefully vulnerable to enemy fighters. Newer, more capable designs became available in time, and a switch to night flying helped to reduce losses. But night missions exacerbated already serious 'frictional' problems associated with navigation and bombing accuracy, with the result that pinpoint raids on individual factories proved impossible. Indeed the 'Butt Report', a 1941 analysis of bombing accuracy, yielded the discouraging conclusion that only one-third of aircraft credited with attacking their targets actually arrived within five miles of them.[24] One result of this finding was to encourage a switch to area bombardment, in the hope that large numbers of bombs scattered over urban targets would destroy some installations that were important to the enemy war effort. Such an approach could not be expected to deliver rapid victory, however, and in the meantime the Germans received plenty of opportunity to refine their defensive techniques against night raids.

A struggle for technical superiority ensued between RAF bombers and German air defences, which sucked in a great deal of blood and treasure on both sides. Temporary fluctuations in this balance could occasionally produce spectacular results. In 1943, for example, the RAF mounted a series of raids against Hamburg, in the process employing improved navigational equipment and new countermeasures against German radar. By such means it succeeded in burning the heart out of the city and temporarily bringing war production to a standstill there. In the immediate aftermath of the raids, Germany's minister for war production, Albert Speer, was moved to warn Hitler 'that a series of attacks of this sort, extended to six more major cities, would bring Germany's armaments production to a total halt.'[25] But the destruction of six more cities in such a manner proved beyond RAF capacity. German defensive tactics were rapidly revised to accommodate the reduced efficiency of their radar, with the result that they were able to punish a subsequent series of major raids against Berlin. Whereas the RAF's average loss rate over Hamburg was just 2.8 per cent of aircraft, over Berlin it almost doubled to 5.2 per cent.[26] Moreover, the damage inflicted on the German capital – while extensive – was nothing like as severe as had been the case in Hamburg. Indeed, until the final stages of the war the RAF was still suffering serious losses in the teeth of German defences without being able to strike a truly crippling blow against the enemy war effort. To be sure, the bombing campaign played an important part in bringing Germany to its knees by destroying its war industry and attracting precious manpower and resources to the air defence mission. But this only really mattered in the context of a wider effort designed to use up the products of that industry, an effort that was well beyond Britain's means alone.

Instead, from mid-1941 it was the Soviet Union that bore the main effort in terms of 'using up' Germany's war production – and indeed her manpower. Having over-run France in spectacular fashion, Hitler was emboldened to tackle the Soviet Union in a similar manner, convinced as he now was of the superiority of German military technique. To a degree he was correct, in the sense that the Red Army was initially ill prepared to cope with the German challenge. But unlike the French, the Soviets were able to trade space for time in the wake of the German onslaught, to learn from their early mistakes and – by dint of a heroic effort – to reconstitute their armed forces along lines that ultimately allowed them to best the Germans. Improved weapons were fielded for employment in accordance with an innovative 'Deep Battle' doctrine, which furnished the Red Army with a set of operational techniques that were, if anything, rather more sophisticated than their German equivalent.[27] Moreover, once it had been shipped east of the Urals, Soviet industry did an excellent job of providing the materiel necessary for a prolonged war of the kind that ensued once the initial German technical advantages had been eroded. And all this effort was sustained by a totalitarian system that meted out a combination of draconian punishment and spiritual inspiration of a kind that galvanized flagging spirits in even the most trying times.

In 1937 Marshal of the Soviet Union Mikhail Tuchachevskii had downplayed the new military techniques, arguing that once

> the Germans meet an opponent who stands up and fights and takes the offensive himself [the] struggle would be bitter and protracted . . . In the final resort, all would depend on who had the greater moral fibre and who at the close of operations disposed of operational reserves in depth.[28]

Victory, in other words, would not stem solely from superior technique: moral and material considerations would also weigh heavily in the balance. In this regard Tuchachevskii was to be proved correct. Once the German armed forces lost their technical edge in the face of qualitative competition from their surviving enemies, there could be no further hope of rapid victories won by small forces at limited cost to themselves. Henceforth, the war's outcome would turn on the quantity, as well as the quality of the armed forces, along with the willingness of nations to endure the sacrifices associated with sustaining an attritional struggle over the long term. In none of these areas did Germany possess an advantage over the alliance of enemies it had created for itself by 1942, which by then also included the United States with its massive industrial potential. Thus, during the following year the numerical balance tipped dramatically against it, and defeat was thereafter only a matter of time.

Bloch's shade

It emerged, therefore, that military technique was not particularly effective at restraining the costs associated with fighting the Second World War. Technique was offset by technique, with the result that the shade of Bloch continued to hover over events. Liddell Hart's ideas for turning this situation to the advantage of liberal powers might have worked under different political conditions, but the fact of the matter was that he failed to appreciate the extent to which political would prevail over technical considerations in the strategic calculations of the belligerents. In this regard he was estranged from the mood of the times: he failed to appreciate the monomania that drove Hitler to throw rational calculation to the wind, and to risk everything in a daring offensive against France. So appalled was he by the prospective costs arising from total war that Liddell Hart could not readily conceive of any political leader thinking differently in this regard.

In its wake, an embarrassed Britain faced little choice but to throw virtually everything she had into her war effort merely to keep Germany at bay. Something approaching total means, in other words, was required to achieve even limited strategic objectives. Quite how long liberal values could have survived inside their island fastness under such conditions is difficult to say. Britain had introduced military conscription shortly before hostilities began, and thereafter set her economy on a remarkably efficient war footing. In the meantime a coalition government kept up the business of parliamentary democracy, while Churchill – who well understood the existential threat represented by Hitler – provided a charismatic brand of war leadership that helped sustain national morale. It soon became apparent, however, that hostilities could not be maintained without an unprecedented degree of additional social and economic direction by government. Attempts to introduce such direction would certainly have proved deeply unpopular, bringing the tensions between prosecuting total war and sustaining liberal values into sharp relief. Ultimately, Britain was spared from the necessity of making existential choices of this kind by the entry of the Soviet Union and the United States into the war. Here were powerful allies committed to Hitler's overthrow and possessed of the manpower and resources necessary to reprieve Britain from its terrible plight. Thus although she emerged at the end of the war with her economy in tatters, Britain nevertheless retained a political system that permitted the 'warlord' Churchill to be voted out of power in 1945.

But if Liddell Hart's strategic prescriptions had been irrelevant to the challenge posed by Nazi Germany, the final few days of the war held out the possibility that they would become rather more apposite in the near future. Writing in response to the atomic bombing of Japan, Liddell Hart argued that new weapons of this magnitude would henceforth demand a great deal more caution in the formulation of strategic objectives than had hitherto been the case. In a situation where

> both sides possess atomic power, '*total* warfare' makes nonsense. Total warfare implies that the aim, the effort, and the degree of violence are unlimited. Victory is pursued without regard to consequences.

It followed from this that an 'unlimited war waged with atomic power would be mutually suicidal.' Consequently even somebody as rash as Hitler would likely be intimidated into pursuing a more prudent, calculating line under these conditions. Technical factors would, in other words, always overshadow political considerations.

Moreover, Liddell Hart was unconvinced that some form of technical counter measure would emerge to change this state of affairs. He acknowledged that the potency of new weapons had hitherto always been offset by countervailing developments, and that 'experience has shown that no development is ever quite so overwhelmingly potent as it appears in anticipation, or even on the promise of its first performance.' Nevertheless, he remained of the opinion that 'atomic energy is such an incalculable force that it is doubtful whether we are justified in reckoning on this bearing from past experience.' In this regard, it hardly needs observing, he was to be proved correct.

Still, none of this necessarily meant that there would be no wars between atomic powers. Rather, such wars as did occur would be conducted in a manner carefully calculated to exploit an adversary's reluctance to initiate a mutually ruinous atomic exchange. Defence against such measures, he argued, would require highly mobile conventional forces capable of responding rapidly to any such emerging challenge and to prevent it from creating a situation that could only be addressed by the use of atomic weaponry.[29] The strategic objective of such forces would not be to disarm an adversary, but to coerce him into ceasing his aggression by persuading him that to continue would ultimately result in his suffering disproportionate costs. In such a manner did Liddell Hart anticipate a basic proposition associated with the limited-war theorists of the nuclear era: if wars were fought at all, it would be necessary to restrict their costs via political as opposed to technical action. But by the time he was elaborating his views on such matters, the main burden of defending liberal democracy was passing to the United States, and it was US ideas on strategy that consequently came to the fore. Thus at this point we leave Liddell Hart behind, and in the following chapters focus on the practice of strategy as it was understood and developed across the Atlantic.

4 The United States and liberal-capitalist war, 1941–1961

When Washington stepped up to guarantee Western security after the Second World War, it accepted a strategic challenge that in certain respects it was poorly prepared to meet. The problem was not one of means, the United States being the most technically advanced and economically powerful nation of its day. Rather it was one of ends, in the sense that US strategists had traditionally eschewed any systematic consideration of the problems associated with using force in pursuit of limited political purposes. Moreover, while they embarked on such a consideration in response to the new strategic challenges posed by the Cold War, they did so only reluctantly. This reluctance stemmed from the historical complexion of US politics. Unlike the French republic, which spent its formative years fighting ideologically opposed neighbours, the United States had enjoyed a far more benign environment in which to invent itself. Its distance from Europe meant that, once it had sloughed off British rule, the new republic did not need to worry overmuch about foreign interventions. The result was a decidedly 'Rousseauesque' constitution: one that privileged individual liberty over state power in the expectation that such a virtuous arrangement would yield correspondingly virtuous politics.

One consequence of this decidedly liberal philosophy was the view that the instrument of war should be reserved for matters that transcended the pursuit of mere state interest. Although the President held the role of commander-in-chief of the armed forces, power to declare war was vested in Congress. This meant that war became a feasible proposition only if it commanded popular support, which in turn demanded that high-minded matters of principle were demonstrably at stake. The sentiment is exemplified by the words of President Franklin D. Roosevelt who in 1941 justified US rearmament in terms calculated to set principle ahead of interests. 'The world order which we seek is the cooperation of free countries, working together in a friendly, civilized society.' This involved 'the supremacy of human rights everywhere [and to] that high concept there can be no end save victory.'[1]

The notion that war was an earnest activity, conducted in support of moral principles, demanded a correspondingly earnest choice of strategic objectives. Since there could 'be no end save victory', it followed that war must be conducted with a view to rendering the enemy defenceless. To stop short of this goal would be to compromise on those same matters of principle that had justified the resort to war in the first instance. The formulation of strategic objectives therefore admitted of no limitation. From this perspective, army chief of staff General George C. Marshall's observation that Clausewitz had 'described war as a special violent form of political action' was meant as a denunciation.[2] Clausewitz might have defended himself with the observation that the US approach to war only *seemed* unpolitical because the goals for which it was fought invariably demanded the formulation of unlimited strategic objectives. But whatever the problems with Marshall's reasoning, his words serve to underline the point that politicized strategies, designed merely with coercion in mind, hardly recommended themselves to the traditional US way of thinking about such matters. This, indeed, is why the Korean War (unlike the Second World War) would subsequently prove such a controversial enterprise: war for 'limited' objectives was popularly viewed as being a compromise with evil.

In short, the traditional US attitude towards war was that if it was worth waging at all, then it was worth waging properly. As such, Alexis de Tocqueville summed up the situation nicely when in 1831 he predicted that 'all the warrior princes who rise up within great democratic nations will find it easier to conquer with their army than to make it live at peace after victory. There are two things that a democratic people will always find it painful to do: to commence war and to end it.'[3]

Since its inception the United States had also entertained equally fixed views on the matter of military means. Generally speaking it was against them; but if they could not be dispensed with, then a citizen army was considered the least objectionable option. Large professional forces were believed a poor fit with liberal values, evoking as they did the twin spectres of militarism and economically unproductive costs. The most that was tolerated in peacetime, therefore, was a small cadre of regulars to act as a nucleus around which civilians could be mobilized to fight when the need arose. The moral qualities of citizen soldiers, imbued with a strong sense of the righteousness of their cause, were expected to offset any technical inadequacies arising from a lack of experience in the arts of war while simultaneously expediting the process of transforming them into efficient fighters.

The American Civil War (1861–1865) demonstrated that there was some truth to these beliefs. Although it is popularly supposed that the terrible casualties of that war were a result of the new rifled muskets with which it was fought, there was rather more to the matter than this. The butcher's bill was really consequent on the willingness of citizen soldiers on both sides to throw themselves into battle after battle in the defence of dearly held values that they considered to be under threat.[4] Moreover, these soldiers proved more than capable of becoming tactically proficient along the way. According to the British military theorist and historian, Colonel G.F.R. Henderson, the 'tactics of the American troops, at a very early period, were superior to those of the Prussians in 1866.'[5] It is therefore not wholly surprising to find that, as late as the First World War, it remained fashionable to argue that newly raised US riflemen, imbued with republican zeal, would suffice to break the deadlock of trench warfare, where their European counterparts had hitherto singularly failed. For his part, General John J. Pershing, the commander-in-chief of First US Army, believed that French and British forces were far too inclined to remain trench-bound and rely on the attritional effects of their artillery to win battles for them. This he considered to be a mistake, because victory demanded that the enemy

> be driven from the trenches and the fighting carried into the open. It is here that the infantryman with his rifle, supported by the machine guns, the tanks, the artillery, the aeroplanes and all auxiliary arms, determines the issue. Through adherence to this principle, the American soldier, taught how to shoot, how to take advantage of the terrain, and how to rely upon hasty entrenchment, shall retain the ability to drive the enemy from his trenches and, by the same tactics, defeat him in the open.[6]

When his soldiers entered the war in 1917 they proceeded to fight in accordance with these beliefs, mounting a series of offensive operations with an alacrity that had long since been bled out of their allies. But although these operations were not without effect on the enemy, they also produced heavy US casualties; and along with these casualties came a more circumspect attitude towards the capacity of citizen soldiers under modern battlefield conditions. Little wonder, then, that when the United States prepared itself to fight a second war on the European continent it sought to identify ways of conserving its precious manpower. This it proposed to do by substituting its abundant capital for soldiers. Liberal warfare was, in other words, to be transformed into liberal-capitalist warfare.

This intention is clearly demonstrated in Marshall's decision to restrict the number of US divisions raised to just 90 during the Second World War, a number that represented a very modest commitment to traditional modes of combat power. One reason for this was that Roosevelt was determined on transforming the United States into what he famously termed the 'arsenal of democracy'. This meant that industry was expected to furnish US forces with everything they required, and also to supply large quantities of weapons and associated equipment to allied powers. In this it succeeded remarkably: not only were US forces lavishly equipped by the standards of the day, but the dollar value of the armaments supplied to allies was sufficient to equip 588 armoured or 2,000 infantry divisions. Maintaining this level of productive capacity did, however, demand the imposition of strict limits on the manpower available for conscription into the armed forces proper. Factory workers could not become soldiers without deleterious consequences for war production.

Moreover, although the armed forces grew to a relatively modest operating strength of 7.7 million men, the vast majority of them were not destined to serve as infantry with rifle in hand. Rather, their roles were to operate and maintain the increasingly numerous and complex weapons that equipped fighting formations based thousands of miles from home.[7] According to Marshall, the United States

> allocated manpower to exploit American technology. Out of our entire military mobilization of 14,000,000 men, the number of infantry troops was less than 1,500,000 Army and Marine.
>
> The remainder of our armed forces, sea, air, and ground, was largely fighting a war of machinery. Counting those engaged in war production there were probably 75 to 80,000,000 Americans directly involved in the prosecution of the war. To technological warfare we devoted 98 percent of our entire effort.[8]

As far as strategic objectives were concerned, all of this new war machinery was intended for use in accordance with a set of innovative doctrinal concepts, designed to disarm an adversary at the earliest opportunity via rapid air and ground offensives. The United States had long had its own group of vocal airpower theorists who preached a broadly similar line to that encountered in interwar Europe.[9] Inspired by these views, the US Army Air Force (USAAF) subscribed to a vision of warfare in which large formations of long-range bombers would conduct raids against carefully chosen industrial targets, whose destruction was expected to impose catastrophic strains on enemy war production. To this end aircraft, such as the new B-17 'Flying Fortress', were equipped with sophisticated bombsights that were intended to provide high levels of accuracy against relatively small targets, at least under conditions of good visibility. Raids were therefore to be conducted in daylight, and the bomber formations were expected to defend themselves against interceptors by flying in close formation so as to create interlocking fields of fire for their generous complement of machine guns.

In the meantime, the US Army had by no means neglected the issue of land warfare. It had studied the German victories in Poland and France very carefully and concluded that the doctrine and equipment of its ground forces should be modelled on much the same lines as those of its future adversary. The function of armoured formations should be to achieve a rapid decision in the enemy's operational depth rather than engaging in an attritional slogging match with his front-line forces.[10] To this end these formations were to be plentifully supplied with tanks, and their supporting arms were to be comprehensively mechanized so that they could play their full role in mobile warfare.

Fighting the Second World War

Just like their British allies, US armed forces soon learned that their ambitious theories would not translate unproblematically into practice, either in the air or on the ground. For their part, daylight bombing techniques did not yield anything like the results that had been anticipated for them. In part this was a result of the frictions of war, not least of these being the weather. Europe was almost always far cloudier than the clear desert sky over the USAAF's bombing ranges where high levels of accuracy were rather more feasible. German air defences also proved much more effective than had been anticipated: anti-aircraft fire proved very dangerous to bombers flying in close formation, while the Luftwaffe developed techniques for breaking up such formations and thereby rendering them more vulnerable to interceptors. Far from dealing a rapid series of crippling blows against Germany's war economy, therefore, the USAAF found itself sucked into a gruelling war of attrition with German air defences.

USAAF fortunes reached a low point in 1943 during its raids on the ball-bearing factories at Schweinfurt. These factories accounted for some fifty per cent of German production, and thus represented a highly attractive target in US eyes: their destruction would drastically reduce the supply of ball-bearings, thereby bringing the German war machine (literally as well as metaphorically) to a grinding halt. On the other hand Schweinfurt lay deep inside Germany, which meant that the bombers had to fly long distances, in the teeth of heavy resistance, in order to reach them. In the event US losses were heavy indeed.[11] In the process the factories were damaged and ball-bearing production was disrupted, and additional raids might well have achieved more serious results.[12] But at this point the USAAF could not afford further losses on the scale it had just suffered and felt compelled to suspend long-range missions. Shortly thereafter, technical developments intervened to save the day. These arrived in the form of long-range escort fighters capable of accompanying the bombers further along the routes to their targets, while doing battle with German interceptors along the way. By such means the Luftwaffe was eventually worn down and destroyed, thereby providing the bombers with more manageable conditions in which to prosecute their raids.

Moreover, just as with aerial bombardment, US experience with the new techniques of land warfare did not match the more optimistic expectations that had been placed on them. The notion that armoured warfare would yield rapid, low-cost victories was quickly dispelled by the effectiveness of German defensive techniques that had been honed in fighting on the Eastern Front. It emerged that armoured formations could only be pushed into the enemy's depth after his defences had first been broken by more traditional operations in which infantry and artillery predominated, and which were often highly attritional in character. General George Patton's famous dash to the Seine proved possible only after German forces in Normandy had been reduced to a shadow of their former selves.

Throwing tanks into such battles served only to multiply the targets available to the enemy. For the most part, therefore, operations in Europe evinced a rather more stolid character, with enemy strength being steadily reduced via the systematic application of quantitatively superior resources.

Ultimately, therefore, the United States discovered that its military technique was inadequate to the task of delivering rapid victory at minimal cost. Against an opponent like Nazi Germany, a sufficient margin of technical superiority was unachievable. Almost until the end of hostilities whatever advantages the United States obtained were eroded by countervailing German developments. The atomic bomb was an important exception in this regard, but it arrived too late to exert any significant influence over the war's conduct. In the event, therefore, sound technique was necessary, without being sufficient, for victory. Under such conditions, the war's outcome turned on the will and capacity to absorb the human and material costs associated with prolonged hostilities.

Nevertheless, in terms of the human costs associated with the Second World War, the United States got off relatively lightly. Casualties amounted to some 201,367 killed and 570,783 wounded. In the immediate wake of hostilities, Marshall was moved to describe these figures as 'staggering', but they were not heavy in comparison with the losses suffered by the other major belligerents.[13] Indeed, the low level of US casualties had less to do with military technique than it did with the Red Army's capacity for fighting in the face of huge losses. For it was on the Eastern Front that the great blood-letting of the war took place, and where the German armed forces were worn down in massive battles of attrition. Measured in these terms, the scale of the US contribution can only be described as modest. While the US Army raised just 90 divisions of its own, the Red Army destroyed or crippled no fewer than 506 German divisions and destroyed another 100 allied divisions. Of the total German losses of 13.6 million, the war in the East accounted for some 10 million of them.[14] For their part, the Soviets also suffered appallingly during the war. Although accurate statistics do not exist, the number of soldiers and civilians killed during the war may have reached 25 million.[15]

Viewed from this perspective, the Red Army might well be described as the 'weapon of mass destruction' with which Germany was ultimately beaten. So long as the Soviet Union was engaged in the business of killing Germans, the US commitment to the war could be a relatively limited one. More specifically, Washington could pursue total strategic and political ends – the destruction of German means of resistance and the toppling of the Nazi regime – without itself having to undergo total mobilization and shoulder a greater proportion of the human costs. After the war, Marshall admitted that it was 'terrifying' to contemplate the situation had the Soviet Union been defeated and the United States been required to fight on without it.[16] Indeed, it is by no means clear that the United States could have fought on with the destruction of Nazism as its goal.

Cold War and containment

The end of the war brought with it a rapid downturn in relations with the Soviet Union. Once the common enemy had been destroyed, old tensions reasserted themselves. Having been a vital partner in the struggle against Nazi Germany, totalitarian communism now emerged as a serious ideological challenge to liberal capitalism. Moreover, Soviet intervention in political developments among its neighbouring states soon convinced Washington that the ideological clash was already working its way through into geopolitical matters.

The question of exactly what US policy should be towards the Soviet Union in these new circumstances was therefore of great importance. On the one hand, efforts to accommodate Soviet demands in the hope that this would lead to some sort of modus vivendi were rejected. Not only did such a policy sound like appeasement – a distinctly unfashionable concept in the wake of the Second World War – but Stalin's regime was also considered to be inherently expansionist, and thus insatiable in terms of its political ambitions. On the other hand, efforts to rid the world of the Soviet Union would require the United States to precipitate another world war. This, too, was a singularly unattractive proposition, not least because such a war would certainly be vastly destructive. The Soviet Union's heroic capacity for resistance could be expected to present the United States with challenges that dwarfed those she had been required to meet during the Second World War. Defeating Nazi Germany had been demanding enough, and that had been with the co-operation of the Red Army. To defeat the Soviet Union with only minimal help from economically and militarily weak European allies was not a task that the United States could conceive of entering into on a voluntary basis, and adding a handful of atomic bombs into the equation did not change matters very much: it was politically feasible to fight such a war only if Stalin were the one to bring it on.

Ultimately, US policy towards the Soviet Union steered a middle course between the two extremes of appeasement and regime change. Observing the situation from the US embassy in Moscow, George F. Kennan famously advocated 'a long-term, patient but firm and vigilant containment of Russian expansive tendencies', and his advice found receptive ears back in Washington.[17] The Soviet Union was, in other words, to be kept in check until such time as it mellowed into something that could be dealt with in a less confrontational manner. In the event, the policy of Containment was successful in navigating between two highly undesirable extremes. It took the best part of half a century to come to fruition, but this is hardly a serious indictment when the alternatives are considered. Along the way, however, it did present the United States with some very serious challenges, not least in relation to the question of how armed force might be used in support of such a policy.

Marshall's views notwithstanding, early thinking on this issue explicitly engaged with Clausewitz's characterization of war as a 'continuation of policy, intermingled with other means.' Nevertheless, the possibility that war with the Soviet Union might be limited to such acts as were strictly necessary for the defence of the status quo was rejected. In the event of war it was anticipated that popular opinion would demand more ambitious political goals, leading to a concomitant escalation in strategic objectives. With this in mind, the United States planned to fight for goals that extended to the complete elimination of Soviet political power beyond the borders of the Russian state. It was believed that attempting to overthrow the Soviet regime outright would be disproportionately costly. Any such efforts were considered likely to provoke fanatical resistance and would also create the requirement to occupy and administer huge territories and populations. The United States therefore contented itself with the prospect that hostilities would have to be terminated by some form of political process. Nevertheless, such an outcome would only be countenanced once Soviet military strength had been reduced to levels at which it no longer constituted a serious threat to its newly liberated neighbours, let alone the world at large. Such a goal necessarily demanded the destruction of the Soviet Union's armed forces along with its industrial capacity for regenerating them.[18]

Essentially, therefore, the United States sought to bolster its limited political goal of Containment with a conditional intention to wage a war characterized by almost unlimited strategic objectives. To this end, early war plans anticipated something that resembled a re-run of the Second World War with the addition of an initial atomic phase. An invasion of Western Europe would be resisted as strongly as possible by conventional US and allied forces, while Soviet economic targets were subjected to atomic bombardment. The United States would then mobilize itself for a longer phase in which it returned to Europe with the goal of disarming the Soviet Union by conventional means.

And then in 1949 the explosion of a Soviet atomic device raised questions about the wisdom of the United States pursuing such ambitious strategic objectives in the event of war. Henceforth, Washington had to take seriously the prospect that atomic bombardment of the Soviet Union would invite a highly damaging response in kind. What this might mean was discussed in an important review of US strategy, which was prepared by the State Department's Paul Nitze. NSC-68 postulated two possible war scenarios. On the one hand, hostilities might open with an atomic attack on the United States, especially if the Soviets had acquired sufficient bombs to mount a powerful blow against key economic and military targets. This would signal the onset of a 'global war of annihilation' in which the United States would seek to render its adversary defenceless using all the means at its disposal. On the other hand, war might begin with an act of local aggression by Soviet conventional forces in pursuit of a limited political objective, while their atomic weapons were held in reserve as a deterrent against US atomic retaliation. Under such circumstances the United States might be better served by refraining from automatically precipitating a global war, and formulating strategic objectives that extended only to combating the local aggression. Otherwise Washington might one day face a miserable choice between two unpalatable alternatives, namely leaving a limited act of aggression unanswered or precipitating an unrestrained atomic war.[19]

Here, in short, was a thoughtful argument for why the United States should consider moderating its traditional approach to waging war. Under conditions of atomic parity, the defence of high-minded principles by means of unrestrained warfare risked incurring terrible costs. If war were to remain a viable tool for containing communist aggression, its conduct would need to be correspondingly limited. Idealism would have to be alloyed with pragmatism, and Clausewitz's politicized conception of war would need to be considered anew.

Although NSC-68 was silent on the details of how limited strategic objectives might be formulated, it had much to say on the forces that were necessary for the prosecution of both local and global war. Strong atomic forces were considered necessary in order to deter a direct attack on the United States, and to strike back if deterrence nevertheless failed. Since such forces would not be decisive in themselves, however, a powerful conventional capability was also necessary in order to hold the ring until the United States could fully mobilize and move onto the counter-offensive. The existence of such conventional forces would also permit local aggression to be met with a proportionate conventional response, rather than an automatic recourse to atomic weapons. Finally, NSC-68 recommended that the United States should push ahead with efforts to produce thermonuclear weapons so as to offset any advantages that the Soviets might gain by acquiring such a capability.

The Korean War, 1950–1953
The political and economic costs associated with creating and maintaining strong conventional and atomic forces, while developing thermonuclear weapons, were such that NSC-68's recommendations would probably have been rejected by President Truman, had not the Korean War intervened in 1950. As it was, the invasion of South Korea galvanized the United States into a crash programme of rearmament. Although Washington had previously excluded South Korea from its list of vital interests in South East Asia, its invasion by the communist North was subsequently felt to demand a robust response both locally and more generally. Thus Truman committed his country to war in Korea and to providing the policy of Containment with a much more serious military dimension.

The opening phases of the war were fought in a traditional enough manner. The North Korean People's Army (NKPA) crossed the 38th Parallel that divided the two states and pushed rapidly southwards against crumbling opposition. Shortly thereafter, the United Nations authorized an armed intervention, and a US-dominated force was established under the generalship of Douglas MacArthur, of Second World War fame. At this stage the NKPA seemed well on the way to victory, but MacArthur succeeded in turning the tables on it by means of a dramatic counterstroke that involved landing forces in the communists' rear at the port of Inchon. Finding its lines of communication suddenly imperilled, the NKPA was forced into precipitate retreat with MacArthur following in hot pursuit.

As operations once again neared the 38th Parallel, it was clear that the status quo ante would shortly be re-established, and it was at this point that questions were raised about what to do next. With communism on the run, here was an ideal opportunity to bring the entire Korean peninsula into the Western orbit. Halting beforehand would leave the enemy free to recover and perhaps cause trouble in the future. With considerations such as these in mind, the United Nations authorized operations to continue north of the 38th Parallel with a view to the 'establishment of a unified, independent and democratic Government in the sovereign state of Korea'.[20] Shortly thereafter, MacArthur was given orders to press on with his offensive and finish the task of destroying the NKPA as a necessary prelude to the project of reunion.

The decision to pursue a military, as opposed to a political, end to the war proved fateful indeed – although not in the manner that the UN had envisaged. Having only just emerged from a prolonged war of its own against US-backed nationalists, the newly minted People's Republic of China was not now about to tolerate sharing a border with a US-backed Korea. Indeed, it was this unpalatable prospect that prompted Mao Tse Tung to intervene on behalf of the struggling NKPA. Accordingly, in late 1950 Chinese troops crossed into Korea and fell upon an unsuspecting MacArthur whose forces were pushed rapidly southwards. The prospect once again loomed of losing the peninsula entirely, although the Chinese offensive eventually ran out of steam as it pushed south, a victim of its own rapidly lengthening lines of communications. This provided MacArthur with a breathing space in which to reorganize along a new front line that approximated to the 38th Parallel. Thereafter superior US military technique severely punished Chinese attempts to renew their offensive.

With the immediate crisis over, new decisions had to be made about the political and strategic direction of the war. MacArthur's command was strong enough to prevent future communist encroachments, but lacked the means necessary to conquer the North. In principle, the option existed of extending the war into China itself by dint of US airpower. Chinese forces would remain an obstacle to the reunification of Korea only in so far as the economic and logistical infrastructure behind them remained intact. Deprived of this infrastructure, they would be unable to resist another ground offensive. MacArthur lobbied hard for an expansion of the war along these lines, convinced that it would not only yield Korea but would also promote the collapse of communist China. Truman, however, refused to sanction such a course of action. China was, after all, allied with the Soviet Union, which meant that either conventional or atomic air strikes against it might provoke Soviet retaliation against Japan or perhaps even Europe. This was a risk that Truman, and indeed his allies, were unwilling to countenance; there was no appetite for initiatives that risked transforming a war over the fate of Korea into a third world war. Thus although it was within the range of US military technique to destroy Chinese resistance on the peninsula, Washington contented itself with the more modest political goal of securing a cessation of hostilities under conditions that closely resembled the status quo ante.

When MacArthur was informed of this he evidently decided to pre-empt any political settlement with an extraordinary public proclamation of his own. In what he later described as a 'routine' communiqué, he pronounced the Chinese incapable of winning in Korea due to their technical inferiority in relation to UN forces. The enemy, he claimed, could not

> maintain and operate even moderate air and naval power, and he cannot provide the essentials for successful ground operations, such as tanks, heavy artillery and other refinements science has introduced into the conduct of military campaigns.

Under such circumstances, he continued, the Chinese advantage in numbers would ultimately avail them nothing. MacArthur finished on what appeared to be a more conciliatory note:

I stand ready at any time to confer in the field with the Commander-in-Chief of the enemy forces in the earnest effort to find any military means whereby realization of the political objectives of the United Nations in Korea, to which no nation may justly take exception, might be accomplished without further bloodshed.[21]

Conciliation cannot, however, have been the desired result because what MacArthur had effectively done was to brand China as a criminal power that was too weak to prosecute the war to a military conclusion. Truman certainly considered that MacArthur was trying to derail a political settlement, and he responded with a controversial riposte of his own. He sacked the general, in the process confirming the primacy of political considerations in this new phase of the war. MacArthur's boots were subsequently filled by a more loyal replacement in the person of General Matthew B. Ridgway.[22]

With MacArthur out of the way, negotiations could now be opened. They would drag on – against the background of continued fighting – for another two years, before an armistice was finally signed that maintained the division of Korea along pre-war lines. In the interim, the strategic objective of UN forces was not to destroy communist resistance on the peninsula, but merely to support negotiations by raising the costs incurred by Chinese intransigence to unacceptably high levels.

A garrison state?

The politicized character of the Korean War was recognized as marking a significant departure from US strategic tradition, and the lessons that were drawn from the experience were various. To some observers, the war demonstrated that force could in practice be used in the limited manner sketched by NSC-68, that rather than responding to a local challenge to Containment by automatically precipitating a 'global war of annihilation' the United States might instead make only that military effort necessary to frustrate such aggressive designs. As William Kaufmann, a prominent advocate of this approach, noted, the strategy that was ultimately adopted in Korea 'served the purpose of inflicting enormous casualties on the Chinese in a way that permitted the enemy to see the process of deterioration setting in, yet gave him the opportunity to contemplate the alternative of a truce.'[23] To the extent that such a strategy could be replicated in other contexts, the United States might be able to move beyond the prospect that future wars would automatically entail atomic bombs falling on its own cities, and thus its threat to employ force in support of Containment would look more credible to the world at large.

From another perspective, however, the lessons of Korea looked rather different. The war had proved costly in terms of blood and treasure, and Truman had ultimately found it difficult to justify these sacrifices to an electorate that viewed victory in terms of evil vanquished rather than the mere preservation of the status quo. Korea, in other words, was not the kind of politicized war that the US people felt comfortable fighting. Moreover, the requirement to prepare for future wars of this kind was not one that was relished. Rearmament had proceeded on an emergency basis following the outbreak of war in 1950, the costs being justified on the basis that communist aggression in South East Asia might well presage the outbreak of global war. But as time wore on without the war expanding, attention increasingly focused on managing the communist threat over what came to be termed the 'long haul'. Korea was now seen as just the first of what was expected to be a succession of communist probes along the boundaries of the 'Free World'.

In this regard, the problem facing Washington was not simply that costly wars conducted in defence of the status quo would be politically unpopular with the US electorate. There were also economic and political risks associated with merely maintaining the military strength necessary to fight such wars over a lengthy period of time. These risks had already found expression in Truman's newly formed National Security Council, which in 1948 had deemed it necessary to

> develop a level of military readiness which can be maintained as long as necessary as a deterrent to Soviet aggression, as indispensable support to our political attitude toward the USSR, as a source of encouragement to nations resisting Soviet political aggression, and as an adequate basis for immediate military commitment and for rapid mobilization should war prove unavoidable.

Having gone this far, it then observed that 'due care must be taken to avoid permanently impairing our economy and the fundamental values and institutions inherent in our way of life.'[24] This caveat reflected popular concerns that powerful armed forces would be expensive to maintain and over the long term would place severe strains on the US economy, resulting in both high levels of taxation and inflation. In policing the frontiers of the Free World the United States might, if it were not careful, spend itself to death. Additionally, many considered that the effects of long-term, extensive preparations for war would prove detrimental to the republican project by eroding civil liberties and militarizing society. It was against this background that army proposals for 'Universal Military Training' were repeatedly rejected by Congress, before being definitively quashed by President Eisenhower who proved particularly sympathetic to the view that a single-minded pursuit of secure frontiers risked slowly but surely transforming the United States into a 'garrison state'.[25]

Massive retaliation, 1953–1961

It was sociologist Harold Lasswell who had coined the phrase 'garrison state' in the context of the crisis-ridden character of international relations during the 1930s. Sustained international tension might, he suggested, lead to the emergence of states where 'the specialist on violence is at the helm, and organized economic and social life is systematically subordinated to the fighting forces.'[26] In this regard Lasswell's garrison state resembled Ludendorff's conception of a totalitarian state geared exclusively to the preparation for, and conduct of, war. But whereas Ludendorff saw little or nothing to regret in such a vision, it represented the very antithesis of liberal values and was thus a disturbing possibility indeed to US ways of thinking.

Certainly the spectre of the garrison state exercised an important influence over the manner in which the Eisenhower administration approached the challenge of containing communism. The key problem – framed in much the same terms as those used during Truman's presidency – was to 'meet the Soviet threat to U.S. security' without 'seriously weakening the U.S. economy or undermining our fundamental values and institutions.'[27] Eisenhower, on the other hand, considered it important to avoid future situations that obliged the United States to counter local communist aggression by repeatedly committing large conventional forces of its own. Permitting Moscow to set the terms of engagement in this manner would, it was argued, place undue strain on US political institutions and economic strength. Indeed, maintaining a robust capacity to fight conventional war for limited strategic objectives would, ironically, demand something that looked uncomfortably like the total mobilization of the United States.

In order to avoid such an eventuality, Eisenhower opted for a technical fix in the form of the thermonuclear bomb. By the mid-1950s Truman's earlier decision to press ahead with the development of a nuclear capability was yielding impressive results in the form of multi-megaton weapons whose explosive power dwarfed that of the previous decade's atomic bombs. Consequently, truly massive damage could now be inflicted on a state even as large as the Soviet Union. With this point in mind the Eisenhower administration embraced nuclear weapons, placing them squarely at the centre of a new strategy. Henceforth, the United States would seek to deter local communist aggression not with the threat of limited conventional war, but with the threat of unlimited nuclear war. The prospect of nuclear weapons falling on Soviet cities would, it was reasoned, dissuade Moscow from causing trouble on the periphery of the 'Free World'. Thus the United States need not bother to maintain strong conventional forces, as they would be surplus to requirements.

Of course nuclear weapons were by no means cheap, but maintaining even a large nuclear arsenal would place less strain on the economy than the conventional forces that would otherwise be necessary. Moreover, nuclear forces could be operated by a small number of military 'technicians', thereby removing the requirement for large-scale military training of the kind that it was feared might contribute towards the militarization of domestic politics. US technique, in other words, meant that capitalism could be preserved, to preserve in its turn, liberal values. With the aid of nuclear weapons, a war for total strategic objectives could be fought on the back of a comparatively limited mobilization of resources and people.

For as long as the United States could credibly threaten to wage unlimited war in support of limited political goals, nuclear weapons would provide an answer to the challenge of bolstering Containment at bearable cost. The problem was that the capacity to make credible threats of this kind would be seriously undermined once the Soviet Union acquired a nuclear capability of its own. In this regard it was unfortunate for Eisenhower that the Soviet Union became a nuclear power some months before his new strategy was publicly enunciated by Secretary of State John Foster Dulles in January 1954. In one of the more controversial speeches of the Cold War, Dulles outlined the dangers associated with maintaining high levels of defence spending over the long haul, before explaining that the administration had decided to address the problem by reinforcing local efforts to contain communist expansion with the 'further deterrent of massive retaliatory power.' Indeed the 'basic decision was to depend primarily upon a great capacity to retaliate, instantly, by means and at places of our choosing', which was code for relying on nuclear weapons.[28]

In the event, the response to Dulles's address – which was quickly dubbed the 'Massive Retaliation' speech – was largely unfavourable, concentrating as it did on the awkward fact that the Soviet Union now possessed nuclear weapons of its own and that it could be expected to use these against the United States in response to a nuclear strike on its own territory. The question that followed from this was whether US nuclear threats could be expected to deter communist aggression under all circumstances. The threat of massive retaliation might reasonably be expected to deter a Soviet attack directly against the United States, but would it really serve to deter more limited challenges to the policy of Containment? According to William Kaufmann:

> the leaders of the Soviet Union and Red China would hardly endow such a doctrine with much credibility. They would see that we have the capability to implement our threat, but they would also observe that, with their own nuclear capability on the rise, our decision to use the weapons of mass destruction would necessarily come only after an agonizing appraisal of costs and risks, as well as of advantages.[29]

Indeed, Moscow might calculate that Washington would refrain from bombing a nuclear-armed Soviet Union in response to a local act of aggression, that when the chips were down it would prefer to preserve US cities intact and cede those of its faraway allies. Reasoning thus, the Soviets might feel empowered to call Washington's bluff, and it would be little consolation to anybody if Washington turned out not to have been bluffing after all.

As the Soviet nuclear arsenal grew in strength, and the consequences of nuclear war looked ever more grave for the United States, the strategy of Massive Retaliation attracted increasing criticism not only from outside the Eisenhower administration, but from within it too. Even Dulles, previously one of the strategy's strongest proponents, changed his position and argued the point forcefully with the President. Essentially, the United States was once again facing the same problem that NSC-68 had highlighted in 1950: lacking a robust capacity for limited conventional war, Washington would either have to back down in the face of local communist aggression or accept the terrible consequences flowing from an unlimited nuclear war. Throughout his years in office, however, Eisenhower remained steadfast: there would be no departure from nuclear reliance. Such conventional forces as were available might be used to nip minor contingencies in the bud, but anything more serious – and particularly anything leading to a clash with Soviet forces – would automatically result in the unleashing of the US nuclear arsenal. Not until the advent of the Kennedy administration in 1961 did the idea that the United States might fight a limited war with the Soviet Union inform a set of fundamental revisions to strategy. Henceforth, the terrible costs associated with fighting unlimited war would weigh much more heavily in the balance as strategy was revised in an attempt to make nuclear weapons into a more credible means of supporting Containment.

In such a manner was Eisenhower's departure from office followed by an official rejection of the view that war should invariably be conducted with the adversary's disarmament in mind. For the advocates of limited war, this development was considered to reflect a new sophistication in attitudes towards war in the nuclear age. 'Out' was the traditional view of war as a moral crusade that necessarily demanded the enemy's disarmament. 'In' was a new view that treated victory as the product of limited, coercive acts of force designed to shape enemy behaviour along lines amenable to US policy, rather than to rid the world of him altogether. Henry Kissinger had famously drawn attention to this distinction a few years earlier.

Because we have thought of war more in moral than in strategic terms, we have identified victory with the physical impotence of the enemy. But while it is true that a power can impose its will by depriving the opponent of the resources for continued resistance, such a course is very costly and not always necessary. The enemy's decision whether to continue the struggle reflects not only the relation of forces, but also the relationship between the cost of continued resistance and the objectives in dispute. Military strength decides the physical contest, but political goals determine the price to be paid and the intensity of the struggle.[30]

Now it looked as though strategy would indeed be liberated from its moral straitjacket in order to accommodate the new political imperatives of the nuclear age.

As we shall see in the next chapter, however, matters were rather more complex than any simple juxtaposition of morals and politics might suggest. Eisenhower had certainly been prone to viewing the Cold War as a struggle between good and evil, but his steadfast position on Massive Retaliation really arose from his personal views about the feasibility of maintaining any form of control over a war with the Soviet Union. Simply put, he was deeply pessimistic on this point whereas his critics were rather more optimistic. At root this difference turned on matters of personal judgement, and as his former critics turned to the business of revising strategy, it emerged that Eisenhower's judgement had not necessarily been as questionable as had previously been considered the case.

5 Limited nuclear war

Surveys of US nuclear strategy habitually open with a well-known quotation from the work of Bernard Brodie, and this chapter is no exception.

> Thus far the chief purpose of our military establishment has been to win wars. From now on its chief purpose must be to avert them. It can have almost no other useful purpose.[1]

Thus wrote Brodie in his 1946 collection of essays, *The Absolute Weapon*. His point was, of course, that the invention of nuclear weapons looked set to make warfare so destructive that the costs of fighting would eclipse any conceivable political benefits. Under such conditions the only rational goal of strategy was to deter war from breaking out in the first place. As predictions go, this was remarkably accurate. Within a decade deterrence had been explicitly placed at the centre of US strategy in the guise of Massive Retaliation. The policy of Containment would be supported by an avowed intention to mount nuclear strikes against the Soviet Union in response to an attack against the United States or its allies. Faced with such an awesome threat, Moscow, it was reasoned, would refrain from causing trouble.

There was, however, nothing inevitable about these events. The United States felt compelled to develop nuclear weapons for fear that failure to do so would hand a decisive technical advantage to the Soviets. But it did not necessarily follow from this that nuclear weapons would become the mainstay of US strategy during the 1950s. They might conceivably have been kept lurking menacingly in the background while the United States bolstered its conventional forces and developed strategies for their use in local wars. As was noted in the previous chapter, it was Eisenhower's concern to contain Soviet expansion without endangering the US economy and its political institutions that initially encouraged him to rely on the deterrent effect of nuclear weapons. In short, there were choices to be made about the balance to be struck between conventional and nuclear forces during the early 1950s, and Eisenhower might have chosen differently had other priorities been felt to demand it. Why, then, did he cleave steadfastly to the strategy of Massive Retaliation as Soviet nuclear forces grew stronger and increasingly threatened the credibility of US deterrence?

Some reasons for Eisenhower's persistence on this point are hinted at in a presidential news conference held in early 1955. During the conference, efforts to draw him out on the precise roles envisaged for nuclear weapons in the event of future war provoked two important points by way of response:

> Now, nothing can be precluded in a military thing. Remember this: when you resort to force as the arbiter of human difficulty, you don't know where you are going; but, generally speaking, if you get deeper and deeper, there is no limit except what is imposed by the limitations of force itself

which was followed somewhat later by:

> Now, war is a political act, so politics – that is, world politics – are just as important in making your decisions as is the character of the weapon you use.[2]

This all sounds rather Clausewitzian, and it was indeed Eisenhower's understanding of the interaction between military and political imperatives that led him to conclude that war would certainly tend to extremes once it had broken out.[3] Even wars that began over the smallest of political disputes possessed the potential to 'escalate' into something much larger as each side strove to avoid defeat at the hands of the other. Moreover, this dynamic would be extremely powerful in a conflict between nuclear-armed adversaries. The potency of nuclear weapons threatened to compress war's duration to such an extent that no opportunity would exist to correct an initial shortfall in terms of military effort. The pressure to commit everything one possessed in an effort to disarm one's adversary would therefore be intense. And to cap it all, the fact that the two superpowers were ideological enemies created a political context that was as likely to encourage extreme military efforts as it was to restrain them. The President therefore had good reasons for believing that there would indeed be 'no limit except what is imposed by the limitations of force itself', and nor was he alone in this regard. The likes of Albert Wohlstetter and James E. King were impressed by what they considered to be the 'delicacy' of deterrence, along with the likelihood of war escaping political control once it began, and they produced some influential analyses in support of such views.[4]

The catastrophic levels of destruction that would be caused by an uninhibited exchange of nuclear blows led Eisenhower to conclude that war could no longer be considered a continuation of politics. It might make sense to *threaten* war in defence of US interests, but if war actually broke out then all was lost. It was in this context that Eisenhower rejected calls for forces and strategies that might be used in contingencies that did not warrant an immediate move to general nuclear war, and stubbornly cleaved to the starkest possible interpretation of Massive Retaliation. To his way of thinking, such capabilities would not so much bolster deterrence as undermine it by providing false hopes that some form of limited war could be fought with the Soviet Union. Thus, not only would they constitute an unnecessary expense but they would also be positively dangerous to peace.

But if force could not actually be used in support of Containment, how then could such an objective be secured in the face of Soviet aggression? Evidently for Eisenhower, the answer was not to rock the international boat, to avoid any occasion for a crisis.[5] If a crisis nevertheless occurred then the appropriate course of action was to prevaricate, to avoid drawing lines in the sand, and to seek a compromise solution that would permit both sides to step back from the brink with honour intact. Indeed it was exactly this philosophy that Eisenhower put into practice when Khrushchev threatened Western access rights to Berlin in 1958. Throughout the ensuing months of crisis, Eisenhower refused to be pinned down to a clearly defined course of action. He reminded the world of the terrible consequences that were certain to stem from nuclear war but he steadfastly refused to explain, either in public or in private, under exactly what conditions he would initiate such a war. Instead he played for time while searching for a negotiated solution that would avoid any appearance of having caved in under pressure. Ultimately Eisenhower achieved his goal by orchestrating high-level talks between the two superpowers, on the back of which Khrushchev was invited to visit the United States. Pleased at the result, and believing that a more conciliatory line might yield US concessions on other matters, Khrushchev eased off the pressure on Berlin and the crisis temporarily abated.

For Eisenhower, therefore, security in the nuclear world derived from political rather than military action. Beyond the act of creating an efficient nuclear force and threatening to use it in the event of war, there was nothing much in the military line to be done. If little of this was read into Eisenhower's unwavering commitment to Massive Retaliation it was because he was essentially a private man, sure of his own judgement, but rather less confident of his chief advisors who evidently did not grasp the fundamental truths of war in the nuclear age. Moreover, it would never have done for the President of the United States to have disavowed the utility of force in the thorough-going manner that a careful exposition of his philosophy would have required. Hence the stubborn passivity with which he met the rising tide of criticism in relation to Massive Retaliation until the end of his presidency in 1961.[6]

Kennedy and McNamara

In that year Khrushchev was once again in belligerent mood, declaring support for wars of 'national liberation' around the globe and renewing his threats in respect of Berlin. In this context, his predilection for trumpeting Soviet advances in nuclear weaponry appeared doubly threatening. These advances would subsequently turn out to be less substantial than they seemed: long-range bombers and missiles did not materialize in anything like the numbers that US intelligence estimates had predicted. Nevertheless, a newly inaugurated President Kennedy set about bolstering the military capacity of the United States to support its policy of Containment. The man charged with achieving this was Kennedy's new defence secretary, Robert McNamara. Kennedy poached McNamara from the Ford Motor Company whose declining postwar fortunes he had helped turn around via the introduction of new management-accounting techniques. As such he can justly be described as a 'numbers' man, and he carried his basic philosophy with him into the Pentagon. Indeed, one of McNamara's key objectives was to ensure that the entire US defence effort was run as efficiently as possible so as to maximize the security bought by each tax dollar. To this end, he pioneered the extensive use of formal cost-benefit analyses as a means of supporting key decisions associated with the acquisition and employment of armed force. The introduction of such techniques proved controversial, particularly when the results they produced flatly contradicted professional military judgement. The 'authority' of numbers proved difficult to challenge by generals who were not versed in the new techniques, however, while McNamara possessed the will as well as the intellect to use them as a means of stripping out what he deemed to be inefficiencies from the defence effort.

But while McNamara quickly acquired the image of an arrogant technocrat (he was soon dubbed the 'human IBM machine'), there was rather more to him than this. Most importantly for present purposes, his obsession with efficiency never blinded him to the key point that the US defence effort was intended to serve the purposes of politics. No matter how efficiently armed force was created and applied, it could only be deemed effective if employed in a manner that furthered the political goals of the United States. The task of translating military efficiency into political effectiveness is of course the task of strategy, and it was to problems of strategy that McNamara gave a great deal of his thought. Chief among these was the question of how to deter limited acts of communist aggression that were conducted from behind the protection of a nuclear shield.[7]

In the late 1950s senior Democrats had added their voices to the steadily mounting criticism of Massive Retaliation, and during the 1960 presidential contest Kennedy himself had observed that war plans designed to destroy the Soviet Union by means of unrestrained nuclear bombardment were irrelevant to the ways in which Moscow was most likely to challenge Containment. A direct attack on the United States he deemed unlikely: what he considered rather more possible were a series of limited moves against of the Free World, no single one being sufficiently threatening to warrant all-out nuclear war, but together amounting to a major shift in the balance of power. The Soviets' 'missile power' he described as a

> shield from behind which they will slowly, but surely, advance – through Sputnik diplomacy, limited brush-fire wars, indirect non-overt aggression, intimidation and subversion, internal revolution, increased prestige or influence, and the vicious blackmail of our allies. The periphery of the Free World will slowly be nibbled away . . . Each such Soviet move will weaken the West; but none will seem sufficiently significant by itself to justify our initiating a nuclear war which might destroy us.[8]

In order to respond effectively to such challenges, what Kennedy considered necessary was not a single all-or-nothing war plan but a strategy that provided the United States with various options in relation to the use of force. Former army chief of staff General Maxwell Taylor coined the phrase 'Flexible Response' to describe a strategic capacity for action 'across the entire spectrum of possible challenge, for coping with anything from general atomic war to infiltrations and aggressions such as threaten Laos and Berlin in 1959.'[9] This was exactly the kind of strategy that Kennedy wanted, and this he charged McNamara with delivering.

Flexible response

McNamara's replacement of Massive Retaliation with a strategy of Flexible Response marked an important point in the development of US strategy. For the first time the traditional focus on total war was to be supplemented by extensive planning and preparation for limited war. Here, in other words, was an official attempt to make war into a genuine continuation of politics by other means. Hitherto, the policy of Containment had been supported by little more than a conditional intention to destroy the Soviet Union by means of unrestrained nuclear bombardment. Under Flexible Response, the United States and its allies would seek merely to 'frustrate' aggression, using the minimum necessary force. Insurgency would be met by counter-insurgency, conventional attack by conventional defence. Nuclear weapons would be used only if it proved impossible to manage the situation with conventional means, or in response to a Soviet nuclear attack. It was hoped that the use of force in such a deliberately restrained manner would bring hostilities to a halt short of a mutually ruinous nuclear exchange, by convincing Moscow that whatever political benefits it anticipated gaining would be outweighed by the costs of pursuing them in the face of demonstrably robust opposition. Confronted with such a situation, it would opt to stop fighting rather than escalating the war to higher, and increasingly damaging, levels of violence. In such a manner, the United States hoped to restore a degree of proportionality between the costs and benefits associated with defeating communist aggression around the world. If there were some reasonable prospect of this being achieved, then war might once again be deemed a rational political act and Washington's commitment to its policy of Containment be deemed more credible.

A shift in strategy of this magnitude demanded major changes in the composition of US armed forces. Flexible Response made no sense if the Soviets were able to mount a disarming nuclear strike against the United States. Under such circumstances Washington might well feel unable to withhold its own nuclear forces during the early stages of hostilities, for to do so would risk losing them to Soviet attack. Thus although the Kennedy administration's basic idea was to de-emphasize the role played by nuclear weapons in support of Containment, this could only be achieved by major investments designed to ensure that the US nuclear arsenal would survive such an attack in a sufficiently intact state to devastate the Soviet Union by way of retaliation. Only if this were possible could the United States afford to exercise deliberate restraint in its use of force during war. To this end great efforts were made to ensure the survivability of its nuclear arsenal by retiring the most vulnerable weapons and accelerating the procurement of more robust alternatives. Short-range B-47 bombers based uncomfortably close to the Soviet Union were phased out in favour of the longer-range B-52; the silo-based 'Minuteman' ballistic missile replaced older unprotected missiles; and McNamara looked very favourably on the new 'Polaris' submarine-launched missile, which remained essentially immune from detection prior to launch. All this was expensive, but the defence secretary's employment of cost-benefit analyses to support the procurement process helped keep the outlay under control. Weapons whose anticipated performance did not merit their expense, such as the 'Skybolt' air-launched ballistic missile, and the Air Force's much-coveted B-70 hypersonic bomber, were either killed off or sidelined.

Of course, even the most survivable nuclear arsenal might only be kept in reserve during wartime if the United States possessed something else to fight with. In order to provide a realistic alternative to early nuclear use, therefore, McNamara needed to find acceptable ways of generating stronger conventional forces. This problem was particularly germane to the NATO area whose importance was considered second only to that of the continental United States itself. The Kennedy administration's Keynesian views on Federal spending combined with the exigencies of the Berlin crisis to free up additional funding for conventional forces. In consequence the number of combat-ready US divisions had increased from 17 to 21 by 1962, while the percentage of mechanized and armoured divisions within the new total had jumped from 18 to 43.[10] In the NATO context, however, the United States could hope only to provide a fraction of the conventional forces necessary to cause significant problems for a Soviet-led invasion; European NATO members would need to provide the great majority. Unfortunately for McNamara the Europeans remained reluctant to follow the US lead in this regard, arguing that Soviet conventional superiority was so great as to render any conceivable increase in defence spending nugatory. Enough in the way of conventional forces existed to provide a 'tripwire' for NATO's nuclear forces, they argued. Anything additional would be superfluous to this mission and therefore unnecessarily costly. Although McNamara consistently argued against such pessimism he did so without real success, with the result that NATO always relied on nuclear weapons to a greater extent than he might have wished. Indeed, it was in this context that he gave serious effort to identifying ways of maintaining political restraint over the conduct of nuclear war.

Counterforce

McNamara's proposed solution to this challenge is generally referred to as the 'counterforce' strategy. As labels go, it is not particularly helpful because counterforce targeting – which is to say attacking Soviet nuclear forces – was already an important ingredient of Massive Retaliation.[11] Early planning for the employment of atomic bombs had focused exclusively on city busting along lines developed during the Second World War, but once the Soviet Union had begun to acquire bombs of its own, Strategic Air Command (SAC) had responded by adding them to its target lists in a bid to protect US cities. Once in the air, Soviet bombers would be very difficult to intercept, but while still on the ground they remained vulnerable to nuclear attack. A single well-placed bomb could destroy an airfield along with all the aircraft on it. Multiple strikes of this kind could theoretically prevent an adversary from mounting any form of nuclear retaliation, rendering him powerless to respond while his cities were destroyed around him. It was always recognized that completely disarming the Soviet Union would demand an impossible level of efficiency from SAC, that the effects of friction would combine to frustrate any such attempt, thereby providing Moscow with ample opportunity to dispatch numerous surviving bombers against US cities.[12] Nevertheless, it was considered preferable to minimize the Soviet capacity for retaliation to the greatest extent possible, as opposed merely to gritting the national teeth in the face of terrible adversity.

But whereas counterforce targeting had previously been considered an integral aspect of Massive Retaliation, McNamara viewed it as an altogether discrete activity within his new strategy of Flexible Response. In other words, he envisioned counterforce occupying its own position on the 'spectrum of violence' somewhere between limited conventional, and unlimited nuclear, warfare. In this he was deeply influenced by ideas developed at the RAND Corporation, an Air Force sponsored think-tank that had been exploring alternatives to the strategy of Massive Retaliation during the 1950s.[13] Those involved in the study of such matters believed that the United States might conceivably use a counterforce strike as a discrete means of defeating a limited act of aggression at bearable cost to itself. The basic idea involved striking at Soviet nuclear forces so as to erode them as far as possible while deliberately refraining from attacking Soviet cities, which would be held as hostages to Moscow's behaviour. Battering its nuclear forces while holding its cities at risk was designed to place Moscow in a situation in which it faced no good choices other than to sue for peace. If all went according to plan, then Moscow would have nothing to gain by expending its surviving weapons against a much stronger US reserve arsenal: to do so would merely shift the balance of surviving nuclear strength further in Washington's favour. On the other hand, dropping them on US cities would achieve nothing other than provoke a much stronger counterblow against Soviet cities. Under such conditions, Washington would enjoy what Herman Kahn called 'escalation dominance', and the rational communist would find himself with no alternative but to cut his losses and accede to the war's termination on terms favourable to the United States.[14]

Although McNamara was never wholly confident that such a strategy would work in practice, he believed it an improvement over the plans for a single massive nuclear strike that had emerged under Eisenhower. Accordingly, he ordered that counterforce targeting be placed at the centre of a major revision of US strategy for nuclear war. The following year he unveiled the new strategy in a speech to NATO ministers, arguing that

> to the extent feasible basic military strategy in general nuclear war should be approached in much the same way that more conventional military operations have been regarded in the past. That is to say, our principal military objectives, in the event of nuclear war stemming from a major attack on the Alliance, should be the destruction of the enemy's military forces while attempting to preserve the fabric as well as the integrity of allied society.[15]

If McNamara had been hoping for a more enthusiastic response than Dulles had received in 1954 he was to be sadly disappointed. Nobody, it appeared, liked the new strategy very much. In European eyes, any initiative that conceivably made war less costly for the two superpowers might encourage Soviet aggression.

To be sure, it would also make the US commitment to defend Europe more credible; but the problem was that virtually *any* such 'limited' war would produce appalling damage and destruction in Europe, even if it left the homelands of the superpowers relatively unscathed. From a European perspective, therefore, the only rational strategy was to threaten an immediate and massive nuclear response to any attack. If deterrence nevertheless failed, then Europe would be little worse off than would be the case under the new US strategy.

Once details of the new strategy emerged into the public domain, the general idea that nuclear war might be subjected to political control was also widely challenged. Hans Morgenthau argued that even if it were practicable to distinguish between the military and non-military elements of an adversary state, the power of nuclear weapons would mean that attacking military targets would produce massive levels of collateral damage. 'For this reason alone,' he concluded, a

> counter-force strategy would be feasible only on the assumption that all military targets were isolated from population centers by the number of miles sufficient to protect the latter from the destructive effects of a nuclear attack upon the former.[16]

Needless to say, this was not a situation that held in either the United States or the Soviet Union, which made it doubtful that either side would readily be able to distinguish between a limited strike on its nuclear forces and an unlimited strike encompassing its cities. And all this was before the likelihood was factored in of weapons going astray and hitting targets that were supposed to be off limits. Given that, under nuclear conditions, command and control would be very difficult, while the time available for making critical conditions would be very short, there would therefore, exist great scope for accidents and misapprehensions to exert a baleful influence on attempts to control escalation. According to Thomas Schelling the 'counterforce campaign would be noisy, likely to disrupt the enemy command structure, and somewhat ambiguous in its target selection as far as the enemy could see.'[17] In short, there was much to suggest that the effects of friction would quickly render attempts to control the course of a limited nuclear war impotent.

To make matters worse, Soviet officials publicly denounced the whole idea of restrained counterforce targeting as an absurdity. The notion of mutually observed restraints governing the conduct of war was, they maintained, wishful thinking; if war broke out between the superpowers it would be fought to the utmost. There was more here than mere doctrinaire posturing. During the early 1960s the Soviet nuclear arsenal was still small and poorly protected in relation to its US counterpart. This meant that Soviet weapons not used during the early stages of hostilities were likely to be destroyed, which in turn suggested that any decision to initiate war on Moscow's part would also be a decision to wage it without restraint. Indeed in private, Soviet thought echoed Eisenhower's views on the impossibility of political restraint. Thus: 'any local war with the participation of nuclear powers will inevitably grow into a global nuclear war as the danger of surprise nuclear strike will constantly be hanging over the armed forces.' Moreover, nuclear war itself would be an unrestrained activity because strikes directed 'against the vital centres of a country, against its economy, its system of state administration, its strategic nuclear forces, and other armed forces is the fastest and most reliable way of achieving victory.'[18] As time went by and Soviet nuclear weapons became better protected, the pressure to use them as rapidly as possible might be expected to exercise less influence on Moscow's formulation of strategy. But while this might be considered a welcome development in Washington it was also the case that the balance-of-force advantages accruing from a US counterforce strike would be correspondingly reduced by such developments. From this perspective, the preconditions for successful political control appeared to exist in tension with those of successful counterforce targeting.

Of course, the United States might conceivably have responded to this situation by further augmenting the quality and quantity of its own forces in a bid to maintain nuclear superiority; but McNamara was loath to contemplate such an eventuality, because he considered it tantamount to writing SAC a blank cheque. As the Soviet nuclear arsenal expanded and became more survivable, so would the United States require ever-more weapons of its own in order to stay in the counterforce business. Over the long term, the costs of attempting to maintain nuclear superiority would grow massively without the United States being any safer as a result. For a defence secretary who was committed to maximizing the security bought by each tax dollar, any such outcome would be intolerable. And then, on top of all this, came the most dangerous crisis of the Cold War.

The Cuban missile crisis

Khrushchev's attempt to safeguard the Cuban revolution by installing nuclear-tipped ballistic missiles on the island resulted in an international crisis that took the two superpowers to the brink of war. As such, the Kennedy administration's experience of managing the crisis exerted a powerful influence over subsequent attitudes towards the use of force under such circumstances. For one thing, in the midst of the crisis Kennedy had publicly rejected the notion that nuclear war would be subjected to political control. 'It shall be the policy of this Nation' he broadcast to the world at large, 'to regard any nuclear missile launched from Cuba against any nation in the Western Hemisphere as an attack by the Soviet Union on the United States, requiring a full retaliatory response upon the Soviet Union.'[19] Whatever theoretical attractions attached to the limited employment of nuclear weapons had evidently failed to recommend themselves in the context of a real crisis.

On the other hand, a flexible response was sought and achieved below the nuclear threshold. With the threat having emerged so close to home, the United States could bring strong conventional forces to bear in a variety of different ways. Air strikes against the Soviet missiles risked provoking Soviet retaliation elsewhere in the world, and were not favoured by Kennedy – at least as an initial recourse. In their stead, comparatively restrained responses – the concentration of powerful forces on the Florida coast combined with a naval blockade of Cuba – emerged as practicable and valuable courses of action. Such initiatives did not materially affect the Soviet position, but they did serve to signal Washington's resolve and to concentrate Kremlin minds on the risks of failing to reach a mutually acceptable resolution to the crisis. Subsequently, a public pledge not to invade Cuba ultimately provided Khrushchev with the pretext he needed to retreat with honour, and the missiles were withdrawn.[20]

Together with the barrage of criticism that greeted his counterforce strategy, the peaceful resolution of the Cuban crisis by means of 'posturing' with conventional forces served to reinforce McNamara's preference for them over nuclear weapons. To be sure, he continued to hold that strong nuclear forces were vital to deter Soviet aggression, but Cuba had underlined the unlikelihood of nuclear war being subjected to political control. It followed from this that Flexible Response was really synonymous with flexible *conventional* response. Nuclear war, if it occurred, would most likely be unrestrained war.

Assured Destruction

All this prompted a further set of revisions to US nuclear strategy, which culminated in something termed 'Assured Destruction'. In its outward appearance, the new strategy appeared to constitute a dramatic departure from the Counterforce approach. Indeed, Washington's avowed focus on attacking Soviet military targets while avoiding cities was now completely reversed. Henceforth, it was announced, Soviet cities would be targeted while military forces would be ignored. If war broke out and escalated to a nuclear exchange, the United States would seek to destroy the Soviet Union as an economic (but not a military) entity.

Reframing nuclear strategy in these terms held a number of attractions for McNamara. If the prospects of exerting political control over the conduct of nuclear operations were as slight as he had come to suspect, it made sense to try and minimize the chance of war occurring in the first instance by emphasizing the awful consequences that could be expected to flow from it. Additionally, a strategy based on targeting cities (as opposed to Soviet forces) could be expected to reduce the chances of the United States becoming embroiled in a costly arms race. SAC would require a relatively modest number of nuclear weapons to wreak terrible damage on the communist world. McNamara judged that maintaining a capacity to kill between 20 and 25 per cent of the Soviet population, and destroy 50 per cent of its industry, would suffice to deter a nuclear attack on the United States. This translated into forces capable of delivering some 400 megatons against the Soviet Union, which was comfortably within existing capabilities.[21] Thus so long as these forces remained well-protected against attack, there would be no real pressure to add to their number over time. However many weapons the Soviets possessed would not be particularly relevant to the job in hand.

On the face of it, Assured Destruction amounted to a rather peculiar strategy; indeed it might reasonably be characterized as distinctly 'unstrategic'. In the event of a major attack on NATO there would be no attempt to achieve a politically meaningful strategic objective at an acceptable cost. Rather, nuclear operations would amount to an irrational strike against the enemy's cities and their inhabitants; for although it might make sense to *threaten* terrible retribution in response to an act of aggression, it would be irrational actually to carry out that threat. What purpose would it serve other than simple vengeance?

Behind the façade of Assured Destruction matters were somewhat different, however. Although Washington was publicly emphasizing city targeting for the twin purposes of deterring the Soviets and managing its own force levels, actual war plans looked rather different. In practice a great many US nuclear warheads were still aimed at Soviet military targets. In the event of nuclear war SAC would seek to execute much the same form of counterforce strike that McNamara had envisaged in 1962, in a desperate attempt to keep Soviet attacks within bearable limits. Soviet cities would not be targeted for their own sake, and only if the war continued beyond the initial counterforce strikes would they be hit. Despite public declarations, therefore, US nuclear strategy did in fact remain relatively unchanged in character. The strategic objectives in the event of nuclear war would be to limit the damage that Soviet forces could inflict on the United States and to exploit the vulnerability of Soviet cities for war-termination purposes.

In this regard it was helpful to McNamara that upward pressure on force levels was significantly reduced by technical developments. These arrived in the form of the new MIRV (multiple independently targetable re-entry vehicle) system, which permitted a dramatic increase in the striking power of US land-based missiles. MIRV-equipped missiles could each deliver three warheads against separate targets with sufficient accuracy for counterforce purposes.[22] Consequently it proved possible to restrain US defence spending at a time when the Soviets were rapidly adding to their own nuclear arsenal: SAC could acquire more warheads without needing more missiles to deliver them. Nevertheless, MIRV did not represent a solution to improvements in the survivability of Soviet weapons. No matter how successfully Soviet land-based forces were attacked, a growing number of submarine-based missiles would always survive to wreak terrible damage on the United States. At base, therefore, the US move to Assured Destruction reflected McNamara's growing conviction that nuclear war could not be a rational tool of politics. Indeed, by the mid-1960s the gap between the Eisenhower and Kennedy administrations on matters of nuclear strategy had greatly narrowed, and Eisenhower's judgement was being to look sounder than had previously been considered.

More options

McNamara left office in 1968, and official statements in favour of Assured Destruction did not long survive him. In 1972 President Nixon gave voice to the key problem popularly associated with the existing strategy.

A simple 'assured destruction' doctrine does not meet our present requirements for a flexible range of strategic options. No President should be left with only one strategic course of action, particularly that of ordering the mass destruction of enemy civilians and facilities . . . We must be able to respond at levels appropriate to the situation. This problem will be the subject of continuing study.[23]

If nothing else, he was certainly correct on the latter point: for the rest of the Cold War the problem of building increased flexibility into nuclear strategy would indeed be a 'subject of continuing study.'

What followed was essentially a reformulation and elaboration of the basic ideas adopted by McNamara during the early 1960s. Publicly this was done in reaction to the shortcomings of a strategy based solely around a threat to bomb Soviet cities. Behind the scenes, however, it was motivated by an increasing dissatisfaction with the counterforce options that were McNamara's lasting legacy. By the 1970s these were beginning to look every bit as unsatisfactory as was declaratory strategy. The size of the Soviet nuclear arsenal was now broadly on a par with its US equivalent, while its survivability in the face of attack was rapidly improving. This meant that any US counterforce strike along the lines envisaged by McNamara would now need to be so powerful and extensive that it would be very difficult to distinguish it from an unrestrained attack on Soviet cities. It would be difficult, in other words, to preserve the cities as hostages while inflicting significant damage on Soviet nuclear forces.

In response to this problem, Nixon's defence secretary, James Schlesinger, was charged with identifying rather more limited ways of employing nuclear weapons, with a view to closing down hostilities without recourse to a full counterforce strike.

Should conflict occur [Schlesinger subsequently explained], the most critical employment objective is to seek early war termination, on terms acceptable to the United States and its allies, at the lowest level of conflict feasible. This objective requires planning a wide range of limited nuclear employment options which could be used in conjunction with supporting political and military measures (including conventional forces) to control escalation.[24]

The basic idea here was to employ nuclear weapons in a manner calculated to frustrate local or otherwise limited acts of aggression, and in the process raise the anticipated costs associated with further aggression to unacceptably high levels. Only if escalation could not be controlled in such a manner would a major counterforce strike be countenanced.

In this regard, Schlesinger's basic approach set the pattern for the rest of the Cold War. Limited nuclear options were central to the Carter administration's 'Countervailing' strategy, which was intended to promote a negotiated end to hostilities by convincing any would-be adversary that 'he would not achieve his war aims and would suffer costs that are unacceptable, or in any event greater than his gains, from having initiated an attack.'[25] Likewise, the Reagan administration saw the limited use of nuclear weapons as an important component in a strategy designed 'to seek earliest termination of hostilities on terms favourable to the United States.'[26] Large counterforce strikes remained an integral part of both the Carter and Reagan strategies, but it was hoped that any war could be brought to an end without resorting to such highly destructive measures.

In the event, however, efforts to integrate limited nuclear options into strategies for fighting nuclear war posed challenges that successive administrations failed to overcome in any particularly convincing manner. The problem that attracted most public attention was that the Soviet Union began 'MIRVing' its own missiles during the 1970s, which threatened to make them substantially more efficient in the counterforce role than they had previously been. This in turn suggested that US nuclear forces would need to be more survivable than hitherto, if the great majority were to be safely held in reserve during the initial phases of a war. It also created pressure for an across-the-board increase in accuracy in order to retain the capacity to strike in a highly discriminating manner as weapons were used up or destroyed in the context of prolonged operations.

One result of this demand for improved accuracy and survivability was the land-based 'Peacekeeper' missile, which became famous chiefly because of the various 'Heath Robinson' schemes proposed for rendering it secure from attack.[27] More satisfactory in this regard was the 'Trident II' missile, whose combination of submarine basing and high accuracy made it a more plausible weapon with which to conduct highly limited nuclear operations. The downside with 'Trident II' was that communicating with submerged submarines was a notoriously complicated and slow process that might rapidly become impossible under nuclear conditions. This, moreover, was only a particularly severe instance of the general problem of maintaining command and control over nuclear operations. The capacity to fight a prolonged and tightly controlled nuclear war demanded command and control systems that were at once highly sophisticated and very robust. Yet there was no realistic prospect that these would be forthcoming. Although the Reagan administration made command and control systems the 'highest priority element' in its ambitious force modernization programme, it was never able to demonstrate a robust capability in this regard.[28] On the contrary, one of the first casualties of even a limited nuclear exchange was likely to be the President's ability to learn what was happening and to set in motion strategically appropriate responses. As had previously been pointed out during the 1960s, the effects of friction always seemed likely to overwhelm any effort to exert political control over nuclear operations.

And if the technical challenges were not daunting enough, the conceptual challenges also resolutely resisted a convincing solution. More specifically, it was always difficult to envisage how a series of limited, coercive strikes would succeed in terminating hostilities on terms favourable to the United States. Each side would retain a plentiful supply of weapons throughout such a war, and it was not obvious why Moscow would capitulate in the face of coercive violence any more readily than Washington. Assuming a rough parity of political will between the two belligerents, such a war might remorselessly escalate into a ghastly slow-motion exchange of cities before political control over operations was finally lost. The only thing that such a strategy would have achieved, therefore, was the destruction of the United States over a period of, perhaps, weeks rather than the day or so it would otherwise have taken. All things considered, this would not be much of an achievement. Indeed, from this perspective, the introduction of limited options into war planning threatened to bring with it a dangerous illusion of control, in the sense that it multiplied existing ways of getting into a war without providing a convincing way of ending it. On this point, Bernard Brodie noted that a good 'way of keeping people out of trouble is to deny them the means for getting into it' in the first instance.[29] This was, of course, exactly the view taken by Eisenhower some years earlier.

The Absolute Weapon

Returning to the title of Brodie's 1946 collection of essays, it would seem that his choice of adjectives was apposite indeed. The power of nuclear weapons was so great as to render it an absolute quality for strategic purposes. Consequently, the pursuit of incremental technical advantages over an opponent was rendered valueless: no matter how efficient SAC became, it could never hope to prevent the Soviet Union from inflicting catastrophic damage on the United States in time of war. On the face of things, this new situation constituted a powerful argument for abandoning the traditional US preference for viewing war in total terms. As there was no technical fix for the unprecedented destructiveness of nuclear warfare, the imposition of political restraints on its conduct would need to be accepted if armed force were to retain any semblance of political instrumentality. It was in this context that Eisenhower's stubborn cleaving to Massive Retaliation led many to question his judgement on matters of strategy.

In the light of subsequent events, however, Eisenhower's judgement appears sounder than many evidently realized. While he was not an expert on the Soviet Union per se, he did possess a shrewd understanding of how a war would likely proceed between *any* two mortal enemies equipped with weapons of unlimited power. Opportunities to exert political control over escalation would have been slight in a war between two ideological opponents whose nuclear weapons remained highly vulnerable until such time as they were used. Under such conditions, war would probably have been fought to the utmost from its outset as each side strove to avoid defeat at the other's hands. In principle, the acquisition of survivable nuclear arsenals by both sides reduced the pressure for, as well as the gains to be expected from, mounting an all-out blow at the very onset of war. In principle, too, this might create a breathing space within which to conduct limited nuclear operations with a coercive purpose in mind. But here once again, Eisenhower's Clausewitzian appreciation of the consequences stemming from the fundamental forces at play in the Cold War – weapons of absolute destructive power coupled to intense ideological competition – provided strong grounds for doubt. Absent a clear asymmetry in resolve, and a war that began with a limited nuclear exchange would remorselessly escalate until the effects of friction intervened to deny any further possibility of political restraint.

It was, of course, always possible for lateral thinkers to draw some comfort from the prospect that the effects of friction would sooner or later confound efforts to maintain control over nuclear operations. In fact, Thomas Schelling famously argued that the real value of nuclear war plans resided not in their practicability, but in their capacity for failure along with the incalculable consequences that would follow. From this perspective, the credibility of US nuclear threats was greater than might otherwise be believed, as it was perfectly possible to imagine that fear, misapprehension or simple accident would produce behaviour in wartime that would never be countenanced under more sober peacetime conditions. Thus, although it might be considered irrational to resort to nuclear retaliation in response to an act of aggression, this by no means meant that such retaliation would not be forthcoming.[30] Still, as Lawrence Freedman has observed, to rely for deterrence on the prospect of matters getting out of hand once the fighting started: '*ce n'est pas la stratégie.*'[31] Quite so, because the practise of strategy involves selecting a course of action from among a range of possibilities, while relying on the prospect of losing control is really to rely on a situation that precludes rational choice. Indeed, to the extent that nobody has succeeded in explaining how a war fought with absolute weapons might reliably yield anything less than absolute destruction, it is questionable whether there has ever been any nuclear strategy.

6 Limited conventional war

Although US efforts to devise a strategy for fighting limited nuclear war were never particularly convincing, the prospect of maintaining political control over the conduct of conventional operations did not seem quite so unlikely. There were both theoretical and historical reasons for believing this to be the case. Theoretically speaking, conventional forces were less potent than their nuclear equivalents, and therefore required much more time to achieve very destructive effects. In principle at least, this provided greater opportunity for political considerations to exert a moderating influence over the conduct of operations. Moreover, the command and control structures necessary to exert such moderation were rather more likely to remain in place than would be the case under nuclear conditions. Historically speaking, the Korean War had already demonstrated the feasibility of using conventional force in a coercive manner in order to contain local communist aggression, while demonstrative deployments had proved valuable in the Cuban context. Whether actual fighting between two nuclear-armed powers could be kept under control remained difficult to say. In the event, however, it was North Vietnam, rather than the Soviet Union, that put the strategy of Flexible Response to the test.

The Vietnam War
Khrushchev's declaration of support for wars of 'national liberation' was interpreted by Kennedy as a serious challenge to the US policy of Containment, and it was in this context that the growing Viet Cong insurgency in South Vietnam became an issue of particular concern in Washington. Kennedy was dead before the United States had become seriously committed to the fate of

Saigon, but his successor, Lyndon Johnson, also considered it important that Containment should not fail in the region. To this end he was willing to use armed force in defence of South Vietnamese sovereignty, albeit within certain limits.

The conventional build-up that Kennedy had initiated in support of Flexible Response provided powerful forces with which to fight in Vietnam: at the peak of the war there would be well over half a million US troops in theatre. The problem facing Johnson was not, therefore, one of means; rather, it was one of ends. More specifically, he considered it necessary to define a set of strategic objectives that would be effective in preserving the South's sovereignty without precipitating direct intervention by China or the Soviet Union. The regime in Hanoi had strong Marxist credentials and thus neither Peking nor Moscow was expected to remain indifferent to its fate. Moreover, the experience of Korea had shown that China was very sensitive to the prospect of US forces encroaching on its borders. All this suggested to Johnson and his civilian advisers that considerable caution was required in the formulation of strategic objectives. Indeed, it was in this context that efforts to disarm the North via a combination of intensive ground and air operations were rejected, not because they were considered militarily infeasible but because they were considered politically undesirable. There was no stomach for turning what was believed to be a proxy war into a direct superpower clash that would bring with it incalculable consequences.

Instead, Johnson opted for a more restrained set of strategic objectives that were intended to coerce Hanoi into reaching a negotiated conclusion to hostilities. To this end, ground forces were deployed to South Vietnam in order to hunt down insurgents and to deal with offensive operations conducted by North Vietnamese regular troops. These ground forces were not to carry the war into the North, however. That was a job for US airpower, whose objectives were to interdict the supply routes that Hanoi used to sustain the Viet Cong, and to bomb various carefully chosen economic targets in the North. The scale and scope of the air offensive was deliberately restrained in the first instance so as to provide subsequent opportunities for increasing the pressure should Hanoi prove intransigent. 'Graduated' increases in this pressure were calculated to bring about an eventual collapse in North Vietnamese resolve without furnishing a serious pretext for Chinese or Soviet intervention at any given point along the way.[1]

But in the event it was Washington, rather than Hanoi, that experienced a failure of resolve. For all its military-technical virtuosity, the United States discovered that it could not bend North Vietnam to its will without incurring costs too high to countenance. In part this was because Hanoi received considerable technical support from China and the Soviet Union. This included supplies of advanced weapons such as fighter aircraft and surface-to-air missiles, along with training in their use. Meanwhile, on the ground, relatively simple weapons were employed in accordance with the techniques of guerrilla warfare so as to minimize opportunities for US armed forces to bring their superior firepower to bear. Nevertheless, all this would probably have proved fruitless had it not ultimately been underpinned by an uncompromising will to prevail. Although the regime in Hanoi was Marxist in character it was also pursuing its own nationalist agenda, and was by no means a Soviet pawn amenable to manipulation in the context of a carefully calculated challenge to US policy. Indeed, while Washington considered itself engaged in a limited war in support of Containment, North Vietnam was stoically fighting a total war in pursuit of reunification with the South.[2] In this context efforts to coerce Hanoi were not going to work.

Matters came to a head with the 1968 Tet Offensive, during which the Viet Cong staged a series of major attacks in South Vietnamese cities, which included an assault on the US embassy in Saigon. For once, these attacks demanded that the guerrillas expose themselves to US firepower, and the Viet Cong was virtually destroyed as a coherent force during the fighting that ensued. Indeed, the commander of US ground forces in Vietnam, General William Westmoreland, evidently believed that he had won an important victory in consequence.[3] But regardless of the military outcome, the political ramifications of the offensive were disastrous for the United States. Until Tet, the Johnson administration had been able to make a case to an increasingly concerned electorate that the war was slowly but steadily being won, that although US forces were suffering heavy casualties they were inflicting worse on their adversaries, who were close to collapsing in consequence. And then Tet vividly demonstrated that this was not the case, that there was plenty of fight left in the enemy, and that the United States was actually nowhere near winning. When, in the wake of the offensive, CBS anchorman Walter Cronkite declared on television that the war had in fact become a costly stalemate, Johnson finally concluded that the game was up, and decided not to run for re-election later that year.[4] In due course the Democrats were defeated by a Republican presidential candidate, who was determined to withdraw US forces from Vietnam.

The chastening experience of defeat in South East Asia prompted a great deal of effort to define exactly what had gone wrong, and how best to address the mistakes that had been made. Proponents of limited war duly pointed out that the character of hostilities in Vietnam had never been amenable to resolution via the kind of highly politicized strategy adopted by Johnson. As Robert Osgood put it, 'the difficulty of conducting a finely tuned campaign of controlled escalation against a belligerent with total aims in a civil war demonstrated the limits of gamesmanship and reciprocal self-restraint in the conduct of limited war.'[5] It followed from this that Johnson and his advisers had erred by failing to appreciate the strength of resolve underpinning Hanoi's bid for unification. As McNamara later explained, this was a result of Washington's 'profound ignorance' of the country and its people. 'I had never visited Indochina, nor did I understand or appreciate its history, language, culture, or values ... When it came to Vietnam, we found ourselves setting policy for a region that was terra incognita.'[6] Consequently the existence was assumed of viable strategic options between the two extremes of disarming Hanoi and staying out of the war altogether, when in fact no such options existed. Vietnam therefore demonstrated the importance of understanding one's adversary as a precursor to effective strategic decision-making. In the absence of such an understanding, even the soundest judgement is likely to be confounded by assumptions that have no basis in reality.

In the event, however, the experience of defeat encouraged a rather different conclusion in many minds, to the effect that there was something *inherently* wrong with the graduated application of force. In his memoirs, Westmoreland argued that 'Although nations in the past have intentionally kept wars limited, as the United States did in Korea, they have applied pressure in terms of their self-imposed restrictions with full force whenever the means were available.' The adoption of a 'graduated response' in Vietnam he considered to be 'one of the most lamentable mistakes of the war' and concluded that victory had really demanded rapid and powerful strikes against key targets in the North from the outset.[7] Rather than slowly ratcheting up the level of violence towards some politically acceptable threshold, he argued, Washington ought to have left the military free to act as it considered appropriate within that threshold.

Clausewitz would have considered this to be questionable analysis. Historically speaking, belligerents have not invariably sought to apply the maximum possible means in pursuit of limited strategic objectives. Rather, the effort expended and the objectives pursued have both been shaped by the importance attached to achieving victory. Moreover, the fact that the United States had lost in Vietnam did not necessarily mean that a graduated approach to the application of force was necessarily doomed to failure under all circumstances. There would always be risks associated with employing limited increments of force, just as there would be risks associated with striking as powerfully as one could manage. A great deal would depend on the particular political and technical conditions of the moment. Nevertheless, in the wake of painful defeat, the idea that the graduated approach was intrinsically wrong, and was therefore to be altogether avoided, found a sympathetic audience within both military and political circles. If the armed forces were anxious to avoid being packed off to far-away places in order to fight with one hand tied behind their collective back, then their political masters were equally unwilling to commit them to such adventures.[8]

Defending Europe

This emerging preference for restricting the use of force to the pursuit of vital political interests understandably focused attention back onto the case of Europe. The security of the United States was intimately bound up with that of its NATO partners, which led many to believe that conventional operations in Europe were less likely to be hedged around with political considerations than elsewhere. To the extent that this was indeed the case, the objective of US armed forces would be '*to win the first battle of the next war*', which in this particular context meant halting the initial phase of a Warsaw Pact offensive by destroying the forces committed to it.[9] But if the question of ends was more to the military's liking than it had been in Vietnam, the question of means was rather less so. Although NATO had formally adopted a strategy of Flexible Response in 1967, European reluctance to spend more on defence ensured that notions of conducting successful conventional operations were as distant as ever. Such forces as existed were stretched thinly along the inner-German border, where they could do little except passively await destruction at the hands of a numerically superior enemy. In this context the US contingent represented little more than a 'hostage' whose early demise, it was supposed, would trigger a nuclear counter blow. By the 1970s, moreover, this situation was looking increasingly problematic because (as we saw in the previous chapter) efforts to identify a practicable strategy for fighting limited nuclear war had not been conspicuously successful. This served to reinforce doubts about the credibility of US commitments to Europe. Would Washington really resort to nuclear weapons, if doing so precipitated a chain of events that would quite probably result in the destruction of the United States? It was therefore considered important to ensure that any war in Europe remained a conventional war, but how could this be achieved without increasing defence expenditure beyond levels that European NATO members were willing to countenance?

One possible solution was to seek to capitalize on a new generation of weapons that seemingly offered spectacular increases in the efficiency of conventional forces. What would subsequently come to be termed 'information technologies' were now beginning to exert a noticeable influence on the character of operations, initially in respect to the accuracy with which targets could be attacked. Towards the end of their involvement in Vietnam, US forces had employed 'precision-guided munitions' (PGMs) to knock out targets such as bridges and armoured vehicles, often with impressive results. The possibilities were even more dramatically highlighted during the October War of 1973 in which Israeli tanks and aircraft unexpectedly came to grief in the face of Egyptian forces wielding a new generation of Soviet guided weapons.[10]

The extent to which PGMs might be expected to transform the character of warfare subsequently became a matter of extensive debate. Their most enthusiastic proponents argued that future battlefields would be saturated with highly accurate fire, making offensive action very costly if not altogether impossible. This was all the more likely, they contended, because PGMs were cheap in comparison with the likes of tanks and aircraft. Intensive investment in the new weapons would therefore greatly improve NATO's capacity to resist a conventional offensive without incurring unmanageable expense in the process. Others dissented from such optimistic forecasts, citing the general principle that military-technical innovation sooner or later elicits counter measures that redress any imbalance, and arguing that this would also be the case in respect of PGMs.[11]

On the whole, this latter school of thought had the better of the argument. It transpired that the setbacks suffered by Israeli forces resulted from the piecemeal and extemporized manner in which they had been thrown into battle during the initial phase of the war. Thereafter it had soon proved possible to devise tactical expedients that greatly reduced the lethality of the PGMs and thus their influence over the course of operations. Indeed, Israeli forces very much had the upper hand by the time a ceasefire came into effect. This was important because, in the European context, the Warsaw Pact could be expected to study NATO's defensive preparations very carefully before attacking, with a view to identifying weaknesses that could be capitalized upon. It would be unwise, therefore, to expect too great a payoff from the new weapons.[12] A better course of action would be to maintain balanced conventional forces, capable of presenting a broader and less predictable range of challenges to an adversary. The problem with such prescriptions was, however, that they ran up against exactly those cost considerations that had made PGMs seem so attractive in the first instance. Thus, although NATO forces did indeed undergo extensive modernization during the 1980s, it remained highly debatable whether such improvements as occurred had furnished the means necessary to 'win the first battle of the next war'.

On a more positive note, buying new weapons was not the only way in which NATO might hope to bolster the efficiency of its conventional forces. It was also possible to envisage improvements in the operational techniques that governed their employment in war. According to this line of argument, the Cold War focus on nuclear matters had so dominated thinking on the use of conventional forces that the traditional 'art' of generalship had been permitted to atrophy. Generals had become accomplished managers of complex peacetime organizations, and some had even developed a sophisticated understanding of the political functions of armed force in the nuclear era. But in the process they had forgotten how to conduct large-scale military operations efficiently. Happily, it was considered, this was an art that could be recaptured via the study of history, and promulgated via new doctrines. Moreover, the costs associated with doing so would be modest in relation to the anticipated results.

It was with these issues in mind that the US armed forces set about revivifying their concern with the more traditional techniques of generalship.[13]

The 'new' vision of conventional warfare that emerged was one that generals of the Second World War would have immediately recognized. Instead of passively awaiting the adversary's blows, emphasis was placed on seizing the initiative from him, on keeping him off balance and forcing him to react to events rather than dictating them. By staying ahead of the game in this manner, it was hoped that US forces would be able to defeat a numerically superior enemy in detail – repeatedly attacking him where he was weak before switching their efforts elsewhere to avoid a concentrated counterblow. Under such conditions a great deal would rest on achieving a relative superiority in the ability to locate the enemy, to understand what he was doing and to formulate a course of action that would pre-empt and frustrate his designs. Success in other words would turn on the ability to cope with the effects of friction more readily than one's adversary.

This new propensity for taking the fight to the enemy was not without its critics. Concerns were raised that any such approach would impose unacceptable escalatory pressures on an adversary equipped with nuclear weapons. As one commentator observed, 'a strong emphasis on taking the initiative, on offensive operations and on winning the battle . . . pays little or no attention to the political limitations on operations which would almost certainly prevail on the European battlefield'.[14] From this perspective, military preoccupations with the efficient employment of force appeared irrelevant to the manner in which operations would actually be conducted, and should not therefore be permitted to exert undue influence on the formulation of strategy. By way of response it was possible to argue that an excessive concern with subordinating the application of force to political considerations amounted to a greater danger. Keeping US forces on a short leash would greatly contribute to their early destruction, thereby hastening the move to nuclear weapons. Conversely, permitting conventional forces greater leeway to act in accordance with traditional operational imperatives increased the prospect of them inflicting a serious reverse on the enemy. This, moreover, could be construed as entirely in accordance with the concerns of Flexible Response under conditions in which it was difficult to imagine how either side would gain from nuclear use. While a serious conventional setback would hardly deprive Moscow of its capacity to defend itself, it would complicate efforts to renew an offensive without recourse to nuclear weapons. Under such conditions NATO attempts to seek a negotiated end to hostilities would proceed from a far greater position of strength than would otherwise be the case. Something of this was captured in the Army's 1986 revision to its basic doctrine. Thus: 'All military operations pursue and are governed by political objectives. Today the translation of success in battle to desired political outcomes is more complicated than ever before.' This was largely due to the fact that 'the risk of nuclear war imposes unprecedented limitations on operational flexibility.' Nevertheless, while such considerations might be expected to influence the pursuit of battlefield success, it remained important to remember that 'defeat will guarantee failure' at the political level.[15]

Ultimately, this latter line of argument proved the more persuasive and European NATO formally committed itself to the new US line. Military efficiency was to be bolstered not only via the acquisition of new weapons, but by the adoption of operational concepts governed by the traditional military preoccupation with disarming one's adversary at the earliest opportunity. In such a manner did the experience of Vietnam influence preparations for the conventional defence of Europe.

The Gulf War, 1990–1991

The extent to which NATO forces would have been let off the leash in the event of an actual war is, of course, impossible to say, the Cold War having ended without the need to answer such questions. However, US forces would soon find themselves conducting major operations elsewhere in the world, which demonstrated that even the most permissive political climate could still, on occasion, impose significant restraints on the application of force.

Saddam Hussein's invasion of Kuwait constituted a flagrant act of aggression against his sovereign neighbour. Moreover, it occurred as the Cold War was ending amidst fervent hopes that international politics would acquire a more benign character than hitherto. Consequently the United Nations proved willing to mandate war for the purposes of liberating Kuwait, and in the process demonstrate to the world at large that aggression could no longer be expected to pay. The United States was by far the largest contributor to the international coalition of forces that was formed for the purpose, and as such became responsible for the strategic direction of the war.

How was Kuwait to be liberated from the clutches of Saddam? Not only was Iraq's army the fourth largest in the world, but it was also understood to be battle-hardened as a result of its recent eight-year war with Iran. Its capacity both to absorb and dish out punishment was therefore expected to be considerable, and the prospect of tangling with it was not taken lightly. Casualties were expected to be heavy, particularly if US forces were drawn into protracted fighting. There was thus agreement between military and political circles that the war should be fought as rapidly as possible to a definitive conclusion, and under such circumstances war planning naturally gravitated towards the strategic objective of destroying Iraq's army at the first opportunity by means of a powerful series of hammer blows. As Chairman of the Joint Chiefs of Staff, General Colin Powell put it, 'First we are going to cut it off, and then we are going to kill it.'[16] To this end, over half a million US troops, along with their associated equipment, were concentrated in Saudi Arabia for Operation Desert Storm. The operation would begin with an air offensive designed to knock out Saddam's command and control infrastructure and impose a high level of attrition on his forces. This would be followed by a ground offensive intended to complete the destruction of Iraqi forces deployed in Kuwait. Thus although Washington's political goal was limited to that of liberating Kuwait, the strategic objective of its armed forces was decidedly sweeping in scope. Moreover, there was a clear determination to employ what Westmoreland had described as 'full force' in pursuit of this objective.

Once the war began, operations proceeded remarkably efficiently. Both in the air and on the ground, supposedly formidable Iraqi forces proved unable to hold their own. Thirty-eight days of bombing, followed by a 100-hour ground offensive, sufficed to render them incapable of organized resistance and to press them into a full-scale retreat. For their part, US combat-related casualties were remarkably low, amounting to some 147 killed and 467 wounded. By way of contrast, Vietnam had cost the United States 47,434 killed in battle along with 303,644 wounded.[17] All this led to the encouraging conclusion that the United States, had succeeded in exorcizing the shade of Vietnam by demonstrating a thoroughly rejuvenated capacity to deliver victory at bearable cost by rapidly disarming its adversary. We've 'kicked the Vietnam syndrome' exclaimed President George Bush Sr in the war's aftermath.[18]

In practice, however, although US armed forces had achieved a remarkable standard of military efficiency, the liberation of Kuwait was hardly down to this alone. Indeed, it was on occasion considered necessary to impose a substantial degree of political control over their operations. Washington was concerned to avoid overstepping the mark in this regard, but did not hesitate to intervene when conditions appeared to warrant it. A notable instance of this occurred in response to Saddam's firing of Scud missiles at Israel, which began shortly after the coalition air offensive got under way. From a military perspective Iraqi Scuds did not represent a serious threat to US forces: they were wildly inaccurate and carried a relatively small conventional warhead. Once they started falling on Israeli cities, however, they became a major political problem. The concern was that if Israel responded by attacking the missile launchers, it might well become impossible for the

Arab members of the coalition to remain in the fight – an eventuality that would in turn have seriously compromised US efforts to prosecute the war to a successful conclusion. Great pains were therefore taken on Israel's behalf in order to forestall such an eventuality. On a daily basis, one third of US air assets were tasked with hunting down and destroying Iraq's mobile scud launchers, and in one instance air strikes were even ordered against a specific list of targets provided by the Israelis to defence secretary Dick Cheney.[19] It is easy to see why theatre commander General Norman Schwarzkopf and his colleagues entertained profound misgivings about these initiatives: not only did they place valuable assets at risk, but they also diverted them from the process of destroying targets that were rather more central to the US strategic objective of disarming Iraq. Nevertheless, Scud hunting was a political priority and so Scud hunting had to be done.

Political considerations also played an important role in bringing the fighting to a halt once it became clear that what remained of the Iraqi armed forces were in full retreat from Kuwait. From a military perspective, the decision to halt operations before destroying these remnants – which included elements of Saddam's elite Republican Guard – was questionable. Such units as survived might, if granted a breathing space, subsequently constitute a threat. Better, therefore, to deal with them while they were still in disarray and incapable of fighting back. From a political perspective, however, the decision to stop was eminently understandable. The goal of liberating Kuwait had been achieved in all but name, and any further fighting would serve no purpose that had been mandated by the United Nations. To continue operations would therefore have been to produce a needless loss of life, and Washington was more impressed by this point than it was by any risks associated with leaving elements of Iraq's armed forces intact.

For all the attention lavished on the military's conduct of operations, therefore, it is clear that matters had greatly benefited from certain timely and well-judged political interventions on Washington's part. Indeed, Powell himself subsequently observed that victory had rested on the ability to formulate strategic objectives that struck an appropriate balance between political and military imperatives. Writing shortly after the war, he concluded that the United States had been successful because 'we ... carefully matched the use of military force to our political objectives.' By the same token, he was also keen to emphasize that 'Decisive means and results are always to be preferred' over gradualism of the kind that characterized operations in Vietnam. This tension between political and military imperatives was not, moreover, something he viewed as unique to conditions in 1991. Rather, it was something that demanded careful management in any war.[20] Here, in other words, we see Powell seeking to articulate the perennial problem associated with the application of force in a strategic context: whereas doing too much risks raising the costs of victory to unacceptable levels, not doing enough provides the enemy with a greater opportunity to strike back, inflicting more damage than would otherwise be the case and perhaps even winning in consequence.

A revolution in military affairs?

Powell's position at the interface between the military and political worlds left him well placed to appreciate that an effective strategy needs to strike an appropriate balance between their frequently competing demands. However, a great many others proved less perceptive in the wake of Desert Storm, concluding instead that superior technique alone had won the day. The whole enterprise had, in fact, been greatly facilitated by the extensive integration of information technologies into the armed forces. This had significantly simplified efforts both to find the enemy and bring highly accurate fire to bear against him. Information technologies had, in other words, succeeded in abolishing a great deal of the friction that had previously accompanied such activities, with the result that US forces had operated far more efficiently and rapidly than their Iraqi counterparts, who were therefore destroyed without the opportunity to respond in kind. It was with such results in mind that Desert Storm was seen as heralding a 'Military Technical Revolution' whose consequences for the future of warfare were expected to be far-reaching indeed.[21]

The notion that US armed forces had stumbled into a technical revolution encouraged a great deal of thought about what kind of efforts should be made to capitalize on the situation. Consciously moving its forces into the 'information age', it was argued, would permit the United States to build on its initial advantages and ensure that its military dominance would henceforth go unchallenged. In the process, the term 'Military Technical Revolution' was dropped in favour of the much broader 'Revolution in Military Affairs' (RMA). The re-branding exercise was intended to underline the point that the challenge went much further than merely buying a new generation of fancy weapons, that exploiting information technologies would mean embracing radical changes to the ways in which the armed forces organized themselves and conducted their operations. In the process, much comfortable tradition would have to be abandoned.[22]

And yet for all this, the RMA's proponents never really succeeded in transcending their original concerns with military technique. At base, claims that US forces needed to pursue a radical programme of doctrinal and organizational innovation were made with a view to improving the efficiency and speed with which destruction could be visited on an enemy. According to one of the most influential proponents of this view – former Vice Chairman of the Joint Chiefs of Staff, Admiral William Owens – the desired goal was to facilitate the delivery of 'devastating firepower' in such a manner as 'to deliver the coup de grace in a single blow.'[23] Contrary voices questioned whether such a capability would be relevant to future wars, whose character might well be different from what was anticipated – not least because the international system was experiencing its greatest upheavals since 1945.[24] Nevertheless, the dominant assumption remained that whatever political obstacles stood in the way of attempting to disarm an adversary at the first opportunity would be overcome by the capacity to employ force ever more efficiently.

In the event, problems associated with treating strategy as a challenge amenable to some form of technical fix would indeed be encountered. The Gulf War had provided something close to ideal conditions for the US military. Having first committed a flagrant act of aggression by invading Kuwait, Saddam had then obligingly lined up his forces for rapid destruction on what amounted to a desert firing-range. In doing so he had generated both a political context that was highly conducive to robust US military action, along with operational conditions in which that action could most easily succeed. Those with an interest in causing trouble in their own particular corners of the 'New World Order' proved well capable of learning from Saddam's mistakes. The fact that it was clearly suicidal to take on the US military at its own game meant that such a course of action could no longer be countenanced; but even if such a military-technical competition were unwinnable, it might still prove possible to derive some important strategic advantages by muddying the political waters to a far greater degree than Saddam had achieved. The trick would be to present the world with a politically ambiguous situation that effectively precluded a decisive military response. By creating a sense that intervention would create more problems than it resolved, it might yet be feasible to achieve gains under the nose of the United States. Some indication of what was possible in this regard would be provided by Slobodan Milosevic's Serbia.

War in the Balkans

Milosevic's efforts to build a greater Serbia along ethnically homogeneous lines presented a particularly serious challenge for US strategy, most notably in respect to the wars in Bosnia (1992–1995) and Kosovo (1998–1999) and their attendant 'ethnic cleansing'. On the one hand, televised scenes of suffering on a scale not witnessed in Western Europe since 1945 created pressure for some sort of military intervention in defence of human rights. On the other hand, the fact that atrocities were committed by all sides served to complicate the goal of identifying exactly what an intervention force might achieve on the ground. Unlike in 1991, there was no prospect of Washington leading an invasion with the object of comprehensively disarming the various protagonists as a prelude to imposing some form of constitutional settlement on the former Yugoslavia. This would have been a massive undertaking for which the political will simply did not exist within NATO. Conversely, notions of intervening in some more limited form held out the worrying prospect of slowly but surely being sucked into a long attritional operation that was likely to involve being shot at from all sides.

All this explains why Colin Powell reacted angrily to a 1992 article in the *New York Times*, which argued for a limited form of military action aimed at protecting Bosnian Muslim civilians from Serbian attack. Senior figures such as Powell were criticized for failing 'to acknowledge that even if collective military intervention cannot readily compel a cease-fire, it can at least slow the slaughter.' Rather than advocating an 'all-or-nothing' approach to warfare, continued the article, he should instead be providing his political masters with a 'range of options' for action in the context of the challenging circumstances presented by Bosnia. For his part, Powell shot back an intemperate reply along the no-more-Vietnams line:

> So you bet I get nervous when so-called experts suggest that all we need is a little surgical bombing or a limited attack. When the desired result isn't obtained, a new set of experts then comes forward with talk of a little escalation. History has not been kind to this approach [and we] have learned the proper lessons of history, even if some journalists have not.[25]

Powell's misgivings about the matter of intervention would subsequently prove to be somewhat wide of the mark in this particular instance. Bosnia was not Vietnam, and there were politically valuable things to be achieved via the limited use of force there. Three years later, when NATO was finally induced to flex its muscles in the face of Serbian intransigence over peace negotiations, it achieved rapid and satisfactory results. A comparatively limited bombing effort, conducted in conjunction with a combined Bosnian and Croat ground offensive, soon brought Milosevic to the negotiating table. Moreover, NATO's considerable margin of technical superiority over Serbia's air defences meant that it lost just one aircraft during the operation.

Nevertheless, the fact that NATO's limited use of airpower had contributed to peace in Bosnia did not mean that Powell's 'nervousness' was entirely groundless. By 1995 the tide of war was running against Serbia, and the pressure to reach a peaceful accommodation sooner rather than later was mounting. NATO was therefore pushing against an opening door, while strong Bosnian and Croat forces were also pushing in the same direction. In this context, the dangers of becoming progressively sucked into some Balkan 'quagmire' were minimal: even a limited application of force was enough to tip matters decisively towards peace. It would have been unwise, therefore, to assume that Milosevic would always readily acquiesce in the face of limited military pressure. The wider context would matter as much as the man himself. And yet this is exactly what NATO did subsequently assume, with problematic results of exactly the kind that Powell had been concerned to avoid.

When in 1998 Milosevic escalated his efforts to bring an increasingly restive Kosovo to heel, NATO proved rather more willing to entertain a forceful response than it had in respect of Bosnia. Misgivings about the limited use of force had been replaced by a sense that Serbia's president was nothing more than a bully. He might prevaricate in the face of threats, but he would back down readily enough when put to the test. Indeed, senior figures in the Clinton administration confidently expected that a few days of bombing would be enough to coerce Milosevic into accepting a constitutional settlement over Kosovo's future. What tended to get overlooked was that Kosovo, unlike Bosnia, was part of Serbia proper. Moreover, it was imbued with great historical importance by Serbian nationalists, and would not therefore be ceded lightly. In this context Milosevic could, and indeed would, resist more stoically than had been anticipated.

For his part, NATO's Joint Air Force Component Commander, US Lieutenant-General Michael Short, was unhappy at the prospect of a limited bombing campaign; although this unhappiness stemmed not so much from his appreciation of the political situation as a desire to avoid repeating what he regarded as the mistakes of Vietnam. As Short subsequently argued:

> when the decision is made to use force, then we need to go in with overwhelming force, quite frankly, extraordinary violence that the speed of it, the lethality of it . . . the weight of it has to make an incredible impression on the adversary, to such a degree that he is stunned and shocked and his people are immediately asking, 'Why in the world are we doing this? If this is just the first night, then what in the world is the rest of it going to be like? How long can we endure it, and more importantly, why are we having to endure it? Let's ask our leaders why this is happening.'

To this end, he considered it necessary to strike as rapidly as possible at the 'strategic target set in Belgrade – the power grid, lines of communication, as they effected [sic] Belgrade the river bridges, the traffic patterns into and out of Belgrade . . . and at least six to eight military command centres in Belgrade.'[26] In this respect, Short's recommendations echoed the approach advocated by the airpower theorists of the interwar period: the adversary was to be rendered defenceless, not by destroying his armed forces per se, but by directly undermining both his will and ability to sustain hostilities.

Short's concern to defeat Serbia as rapidly as possible was motivated by the eminently sensible military reason that the sooner an adversary is beaten the less opportunity he will have to strike back at one's own forces. But if military considerations called for a rapid and uncompromising series of strikes against a wide range of targets, political considerations counselled very much the opposite approach. In particular, the prospect that bombing would produce large numbers of civilian casualties demanded that restraint be exercised over the use of force. Such views were most vocally expressed in Europe, but similar sensitivities existed higher up the US chain of command. According to NATO's supreme commander, General Wesley Clark:

> Once we moved past the obvious air defence target set, every target . . . was in one way or another, likely to become controversial. In the U.S. channel, we would need a complete analysis of each individual target – location, military impact, possible personnel casualties, possible collateral damages, risks if the weapons missed the target and so forth . . . And this had to be done to my satisfaction, then sent to Washington where it underwent additional levels of legal and military review and finally ended up on President Clinton's desk for his approval.[27]

The fact of the matter was that, at the levels that counted, on neither side of the Atlantic was there any great enthusiasm for giving the military free rein to defeat Serbia in the manner it thought best. Political considerations reigned supreme, and only when it became clear that Milosevic was not going to capitulate as readily as had been envisaged was the range of targets expanded to include those of the type contained in Short's 'wish list'. This process of expansion was, moreover, something that occurred gradually in deference to the political sensitivities attending the war.

Short's ambition to reap the maximum military benefit from a rapid bombing campaign was therefore never realized. In this instance, however, his concern to avoid a graduated approach to the application of force was unnecessary. This was because, unlike North Vietnam, Serbia lacked the military technique necessary to contend with US air power. Serbian air-defence systems survived the war, not because they prevailed over their adversaries, but because they were deliberately withheld from combat and hidden away. From time to time a lucky pot shot succeeded in bringing down a NATO aircraft, but there was nothing to suggest that

Serbian air defences became more dangerous as hostilities dragged on. They remained a threat to NATO throughout the war, but ironically their threat-value remained intact only so long as they were never used in earnest.

Indeed Serbia's most effective weapon against NATO was its propaganda machine. By hunkering down under the bombing offensive, and refraining from mounting anything that might be considered an effective military reply, Milosevic hoped to achieve 'victim' status in the eyes of the international community.[28] To this end his propaganda proved rather effective. Bombs that missed their intended targets, or otherwise inflicted collateral damage, were a particular boon in this regard. Indeed, television images of almost *any* bomb damage were amenable to interpretation by quick-thinking, creative minds – the more so because NATO was initially slow to appreciate the importance of providing its own version of events in a timely and robust fashion. Kosovo was, in fact, a 'postmodern' war, in the sense that the stories told about bomb damage were in many respects as important as the actual physical effects it created. Thus if NATO was going to win, it would have to do so not only through accurate bombing but also through telling convincing stories about what was happening and why.

Matters were helped in this regard by the fact that Serbia's campaign of ethnic cleansing in Kosovo proved so unsavoury as to undermine the victim status that Belgrade was seeking to create with the help of NATO bombs. Thus although the air campaign did attract widespread criticism, many were prone to see it as the lesser of two evils.[29] Conversely, efforts to prosecute hostilities to an acceptable conclusion were hampered by the fact that nobody was clear about exactly what combination of bombing and threats would coerce Milosevic into submission. The steady expansion of authorized targets was therefore something of a trial-and-error enterprise, with all the risks that this entailed. As it happened, NATO military technique was good enough not only to preclude casualties among its aircrews but also – despite some unfortunate accidents – to keep instances of collateral damage within politically manageable limits. The pressure on Belgrade could therefore be kept up, although it was by no means clear where events would eventually lead as the campaign dragged on.

Milosevic's decision to capitulate after 78 days of bombing was therefore greeted with a sense of considerable surprise and relief. Surprise because there had been little to suggest immediately beforehand that matters were coming to a head in Belgrade; and relief because it was becoming increasingly difficult to identify any further targets to attack. Thereafter, relief gave way to uncertainty over exactly why Milosevic had surrendered. Which aspects of the bombing had proved effective, and how important were these in relation to the various other military, economic and diplomatic pressures that had been brought to bear on Serbia? In a post-war report to Congress, Secretary of Defense William Cohen and Chairman of the Joint Chiefs of Staff Henry Shelton surveyed a range of such factors, only to conclude that they 'all . . . played important roles in the settlement of the crisis.'[30] Perhaps it is not too unkind to suggest that such a conclusion betrays a failure to get at the root cause of Milosevic's capitulation, and to understand the role of NATO's bombing therein.[31] It certainly reinforces a sense that no matter how efficiently NATO was capable of applying force, its capacity to do so effectively was undermined by a poor understanding of the beliefs and values that Milosevic brought to the struggle.

The 'big battle' philosophy
According to General Wesley Clark, in the case of limited wars such as that fought over Kosovo

> the 'big battle' philosophy that dominated much of Western military thought during the twentieth century must be modified. While nations have always aimed in war to gain their objectives with the least cost, in modern war, achieving decisive political aims may not require achieving decisive military

results. Wars can be won through battles never fought, as much as through the 'battles of annihilation' taught in the military textbooks.[32]

Victory, in other words, depended less on the ability to visit rapid destruction on Serbia than on the capacity to generate political leverage from threats of destruction. Coercive strategies of this kind therefore require sound judgement, supported by an understanding of one's adversary, in addition to technically sophisticated inputs.

During the second half of the twentieth century, the United States excelled on the technical front, as was amply demonstrated by the 'big battle' conducted against Iraqi forces in 1991. On the political front, however, matters were less satisfactory. Desert Storm provided an excellent showcase for US military technique not simply because the Iraqis proved incompetent by comparison, but because the political context of the war was unusually conducive to robust military action. In contrast, Vietnam and Kosovo provided rather more demanding tests of Washington's judgement and found it wanting, not least because nobody really understood what was at stake for the enemy and what would have to be done in order to win. Thus the United States found itself in an impossible situation in Vietnam, unwilling to run the risks and bear the costs that victory demanded despite enjoying a very significant technical edge over its adversary. In Kosovo it proved possible to extract a victory from the war despite the fact that Serbian resolve had been underestimated. In this instance, NATO's technical superiority proved important in keeping the unanticipated costs and risks of the war within manageable bounds. And yet things might well have gone differently if Milosevic had demonstrated a little more resolve or proved able to play the victim card a little more effectively.

But if the challenges associated with formulating coercive strategies were attracting increased attention in 1999, they were very soon eclipsed by other concerns. September 2001 saw the United

States come under direct attack for the first time in sixty years, and the resulting climate was not one that encouraged notions of carefully calibrating the application of force to a complex political situation. A mortal enemy had emerged, who evidently required to be destroyed as rapidly as possible, and 'battles of annihilation' were back on the agenda. The strategic challenges raised by such battles were, therefore, expected to be technical rather than political in character.

7 The 'Global War on Terror'

Al-Qaeda's attacks on the United States in September 2001 convinced the Bush Administration that it was facing a dangerous new enemy whose actions were limited only by the means available to it. The deliberate targeting of civilians was interpreted as a terroristic strategy concocted by fanatics who would stop at nothing to erode political support for US engagement in the Middle East. This was viewed as the first step in a bid to sweep away the existing states in the region, and establish a pan-Islamic caliphate in their stead. Washington's response was to declare a 'Global War on Terror' whose avowed aim was to make the United States safe from a repetition of the 2001 attacks. This in turn was deemed to demand the destruction of al-Qaeda, along with every other terrorist group that harboured international pretensions.

Armed force had hitherto played a relatively minor role in Western responses to the challenges posed by terrorism. The tendency had been to treat it as a criminal act that demanded a police, rather than a military, response. The notion of going to war against groups such as al-Qaeda was therefore a departure from the norm. In this regard it reflected Washington's concern that Islamist terrorists were only part of the problem. Left to their own devices they were already dangerous enough, but with help from 'rogue' states their capacity for causing death and destruction might well be greatly magnified. The nightmare scenario was one in which Osama bin Laden was provided with nuclear weapons that he would use to cause unprecedented damage to the United States. Al-Qaeda had already demonstrated a lack of respect for the convention of non-combatant immunity in pursuit of its political goals, and it was therefore assumed that it would not shrink from breaking the convention relating to the non-use of weapons of mass destruction.[1] As the US *National Security Strategy* of 2002 put it: 'The gravest danger our Nation faces lies at the crossroads of radicalism and technology.' It followed from this that the only way to guard against such an eventuality was to ensure that no state that might wish to do so was ever in a position to provide al-Qaeda with such weapons. Thus the targets of the War on Terror would not only be international terrorists, but also those regimes that were willing to help them achieve their goals. Both 'terrorists and tyrants' were considered fair game in this regard.[2]

Notice therefore went out that Washington expected the co-operation of states within whose borders such terrorist were believed to be lurking. These states must either participate in the hunt or permit US forces to operate in their territory. Refusal to do so would be considered a potentially hostile act. As President George Bush Jr summed up the prevailing mood in Washington: 'Either you are with us or you are with the terrorists.'[3] It followed from this that states giving grounds for suspecting that they were actually aiding and abetting terrorists would be in grave danger of attack themselves; and to brand one-self a problem of this magnitude would be to risk more than a mere slap on the wrist: it would be to gamble with the very future of the regime itself.

In practice, the Bush administration was proposing to deprive al-Qaeda of hiding places and sources of support by accelerating the transition of the world's more politically problematic corners towards a democratic capitalist future. By any reasonable stretch of the imagination this was an ambitious project. Nevertheless, it was one that was considered warranted by the significant advantages the United States possessed in terms of military technique over any likely adversary. A decade of investment since the 1991 Gulf War had produced further efficiency gains among the armed forces. These were the fruits of an on-going programme designed to ensure that, henceforth, the US military would always be committed to battle with a 'frictional imbalance' working to their advantage.[4] This was understood to demand the realization of an information-based Revolution in Military Affairs, which the Pentagon was now undertaking in the guise of its 'Force Transformation' initiative with a view to fielding a 'smaller, more lethal and nimble joint force capable of swiftly defeating an adversary.'[5] Success in this regard was intended to ensure that the costs associated with resorting to war would always remain within politically acceptable levels, even in the face of desperate opponents who were struggling to save their regime from extinction.

The prospect of re-engineering the world's trouble spots in such a manner was also made more palatable by a Rousseauesque conviction that people are naturally benign and well-disposed towards their neighbours, and that problems of international relations are a function of 'evil' states rather than the nations they rule. Always present in US political thought, for the Neoconservatives who surrounded the president it appeared to have been elevated to an article of faith that became a cornerstone of foreign policy. All one had to do was to topple the bad regimes and supply a little economic aid, and liberal capitalism would spring up more or less naturally in their wake. History was, after all, on the side of the United States in this regard, and with powerful allies like this to hand there could be little need to become heavily involved in rebuilding any states that got broken along the way.[6]

In such a manner, it was anticipated, Washington would be able to drain away the sea in which the terrorist fish swam. If this in itself did not lead to the extinction of al-Qaeda, then it would be a relatively easy task to bomb its remaining members into extinction. Thus to the extent that all this was true, the War on Terror (and tyrants) could readily be contemplated as a technical exercise in which the United States held all the advantages. Eliminating threats was the order of the day, and this was a task that could be achieved at reasonable cost through the exercise of greatly superior technique. Achieving this would more or less automatically create political conditions in which terrorist groups would be unable to survive, let alone operate effectively.

Afghanistan

It was only to be expected that the United States would strike first at Afghanistan. It was well known that the Taliban regime had been playing host to bin Laden and his associates in exchange for generous financing. Afghanistan was therefore the first state to be put to the test: would it renounce its support for al-Qaeda and hand over its leadership into US custody, or would it face the consequences of denying Washington's demands?[7] In the event the Taliban leadership refused to comply. In Washington's eyes, this placed them fairly and squarely in the 'tyrant' category, and the US military began bombing in early October 2001. The strategic objective was both to destroy al-Qaeda and to disarm the Taliban.

In one sense Afghanistan represented a considerable challenge to US strategists. For the most part, Taliban and al-Qaeda forces consisted of lightly armed infantry who operated independently of the kind of strategic infrastructure that the air force could readily bomb, and who could easily disperse in the face of direct air attack. That said, Washington was fortunate in possessing an indigenous ally in the form of the Northern Alliance, a disparate collection of Afghan groups, which had banded together during the mid-1990s in order to fight the Taliban, and which with US military support could now be made to render valuable service. Ground offensives by Northern Alliance forces were mounted in order to compel the enemy to concentrate in response. Once concentrated they were highly vulnerable to air attack and could more readily be destroyed in consequence. With the way prepared for them in this manner, Northern Alliance forces were then able to advance and threaten new territory. In the event, the better-trained elements of the Taliban and al-Qaeda fighters proved capable of employing effective concealment techniques in a bid to protect themselves against air attack.[8] Nevertheless, there were limits imposed on such countermeasures by the requirement to defend the approach routes to Afghanistan's cities against Northern Alliance advances. And so the process of 'fire and movement' was repeated until Kabul was captured in mid-November and the Taliban were pushed back onto their spiritual birthplace, the city of Kandahar. By early December Kandahar itself was coming under increased pressure and the Taliban leader, Mullah Mohammed Omar, abandoned it for the mountains, although not before promising to prolong the war in the form of an insurgency. For his part, bin Laden eventually slipped across the border into neighbouring Pakistan. Operations designed to finish off both Taliban and al-Qaeda forces still present in Afghanistan continued into March 2003, at which point victory was declared.

Along the way, the war had produced some unlikely combinations of technical means, which extended to strikes with precision-guided munitions preparing the way for rather more old-fashioned cavalry charges. Nevertheless, all this local colour stemmed from the pursuit of traditional strategic objectives, albeit under an unusual set of conditions. Enemy forces were attacked and destroyed as rapidly as possible with a view to disarming the Taliban regime and destroying al-Qaeda at the lowest feasible cost. The co-option of Northern Alliance forces into the process produced the added bonus of drastically reducing the number of US personnel that needed to be placed within range of enemy Kalashnikovs. By March 2003 US combat-related deaths had yet to reach double figures.

Once this initial phase of the war was over, however, a new and rather more intractable challenge emerged. Hitherto, Washington had focused on achieving its strategic goals at the expense of worrying overmuch about longer-term political questions. Moreover, there was a pronounced reluctance to let US troops become involved in any subsequent reconstruction mission. Once the main phase of the war was over, they were expected to return home as soon as possible. In practice, however, it soon became evident that their presence in theatre would remain essential for some indeterminate time to come. A UN-mandated International Security and Assistance Force (ISAF) was established in December 2001, with the task of providing security in Kabul for the fledgling political institutions that were intended to move Afghanistan into a democratic future. It was soon realized, however, that ISAF would need to do rather more than protect Kabul, that its mandate would in fact need to extend much more widely, and that the United States would have to provide a great many of the additional troops that were needed. The problem was Mullah Omar's promised insurgency, which really began to make its presence felt in 2003. Relieved of the requirement to defend territory, Taliban fighters were now able to adopt a combination of terrorist and guerrilla techniques for prolonging the war in Afghanistan that were far better suited to their conventional military inferiority. By dispersing and avoiding contact under any but the most favourable circumstances, they were able to minimize the effects of their adversary's superior firepower. They would concentrate to attack military or civilian targets with the benefit of surprise, only to melt away once more before a serious retaliatory blow could be mounted against them. Indeed, any such blows were as likely to fall on the heads of innocent Afghans as they were on those of the insurgents – an outcome that was positively harmful to ISAF's cause. Over time, therefore, the effect of the insurgency was to render the reconstruction effort more difficult and expensive, and to erode popular faith in the new government's ability to bring security and prosperity to Afghanistan. In its turn, waning popular support for the government made it harder for its security forces and those of its ISAF allies to operate effectively, and correspondingly easier for the insurgents to step up their own activities. In the longer term, the prospect was one of the international coalition suffering increasing casualties in an effort to salvage a deteriorating situation, before finally deciding to cut its losses and leave the Afghan government to its own fate. It therefore behoved the United States to redouble its efforts to defeat the insurgency before it became a more serious threat. But at the time when this might still have been feasible, Washington was already setting its sights on other targets and withdrawing forces from Afghanistan in order to open up another front in its War on Terror.

Iraq again

By 2001 Iraq had been a thorn in Washington's side for a decade. Saddam's armed forces had been battered into submission in 1991 and no longer represented a serious threat to regional stability. On the other hand the Iraqi dictator remained in bellicose mood, and his refusal to co-operate fully with UN weapons inspectors led many to suspect the existence of a covert nuclear programme that would ultimately permit him to foment new forms of trouble abroad. Moreover, immediately after the 2001 attacks on the United States, the problem of Iraq was perceived in a very different light in Washington. The possibility that al-Qaeda would provide Saddam with the strategic reach he required to attack the United States, by acting as the delivery system for Iraqi nuclear weapons, was taken seriously indeed. Even as it was invading Afghanistan, therefore, the Bush administration was reassessing its position in relation to Saddam and drawing up plans for toppling him from power. To do so, it was reasoned, would not only nip a nascent threat in the bud: it would also reinforce the message that the War on Terror would not be a merely reactive affair. The point would be well and truly rammed home that the United States would not limit itself to retaliatory action in the wake of attacks, but was committed to a policy of eliminating threats. Saddam, it was concluded, had to go.

Removing Saddam from power demanded that US forces be sent to Iraq with the strategic objective of disarming his regime. If he could be accurately located in the interim, then a successful 'decapitation' attack might render a more conventional clash of armed forces nugatory. The focus of military planning was, however, the destruction of Iraq's means of resistance. This focus on the strategic objective of disarming Saddam was hardly objectionable in Washington, where it was accepted that it would be the necessary outcome short of an early collapse of organized Iraqi resistance. Where political considerations did intervene to shape the character of the war was in the size of force that theatre commander General Tommy Franks was allocated in order to achieve his objective. Secretary of Defense Donald Rumsfeld was determined to ensure that nothing more in the way of resources was committed to the enterprise than was absolutely necessary. In part this reflected his faith in the ability of superior US military technique to deliver results without requiring the commitment of massive numbers of troops.[9] His generals, he believed, remained too traditional in outlook and unwilling to reap the benefits associated with the Force Transformation programme that he himself had been championing. Moreover, army views on troop levels included a provision for post-war occupation duties that Rumsfeld did not believe would be necessary. The creation of a new Iraqi state sympathetic to US interests was, he believed, something that would be undertaken by the grateful Iraqis themselves. Thus, for the famously self-confident defence secretary, the whole enterprise was one that could be successfully conducted by relatively modest forces that should expect to be leaving Iraq soon after Saddam had been removed from power. For his part, Franks calculated that he would have a quarter of a million troops on the ground by the time that Iraq was defeated, but that not all of these would actually reach theatre if resistance collapsed especially quickly, and that forces already in position would be rapidly reduced thereafter.[10] There was a certain amount of military concern over the matter of troop numbers, but Rumsfeld had his way. The enemy, on the other hand, would prove to be rather more recalcitrant than had been anticipated.

Operation Iraqi Freedom began on 20 March 2003. Within just three weeks the regular elements of the Iraqi armed forces had either scattered or been destroyed, US forces were in Baghdad and Saddam was on the run. In some respects, therefore, Rumsfeld's belief in the ability of superior military technique to deliver results on the cheap had been justified. Concentrations of regular Iraqi troops and heavy weapons had, for the most part, been summarily swept aside wherever they had been encountered. In other respects, however, the initial stages of the war had not gone in quite the smooth manner that the more optimistically inclined had anticipated. As US forces moved north from Kuwait the most problematic foe they encountered were the numerous *Fedayeen* paramilitaries that dogged their steps. Indifferently trained and equipped only with light weapons, they could not hope to prevail in a stand-up fight. They proved rather more effective, however, against softer targets such as the logistics troops that kept US combat units fighting. As these combat units advanced north-wards they found the threat emerging in their rear increasingly difficult to deal with, and ever more effort had to be diverted to the task of defending their lengthening lines of supply against attack.

Ultimately, the *Fedayeen* could not prevent US forces from reaching Baghdad and forcing Saddam into hiding. On the other hand their 'irregular' exploits presaged the unfortunate shape of things to come. The Neoconservative assumption that an Iraq without Saddam would rapidly transmogrify into a capitalist democracy proved sadly in error. Even had the institutions of democracy been readily understandable by the Iraqi people, there remained enough die-hard paramilitaries in the country to guarantee the emergence of an insurgency that was sufficiently violent to disrupt efforts to build a new state. US forces took steps to combat this growing disruption, but with little success. Although there had been enough of them to topple the Iraqi state, their numbers were insufficient to provide an alternative source of security for the beleaguered Iraqi people. Franks' expectations notwithstanding, in May 2003 there were just 150,000 US troops in theatre and this was a number that would shrink in months to come.[11] In this regard army dis-quiet over numbers had proved rather more sound than had Rums-feld's optimism, as the result was a power vacuum in the country that multiplied opportunities for those with a vested interest in ensuring that a new Washington-sponsored state would not be successful. The original Ba'athist insurgency was supplemented by a sectarian struggle between Sunni and Shia Muslims that had previously been kept in check by Saddam, while al-Qaeda employed its characteristically brutal methods to fan the flames of what threatened to become a civil war, and generally to make it as costly as possible for US troops to operate in Iraq. There was, in other words, rather more of Hobbes than Rousseau to post-Saddam Iraq; and while the violence was certainly 'nasty and brutish' in character, there was no sign of it being 'short' in duration.

Not only were US troops too few in number to deal with the steadily deteriorating situation, but they were also inadequately trained for the job. It is hardly surprising, therefore, that their initial reactions were at best marginally effective and at worst tended to exacerbate the problems they faced. Schooled in the tradition that force is best applied with a view to disarming one's adversary as rapidly as possible, they sought to operationalize this basic approach, albeit on a smaller scale, in Baghdad and other urban areas. To this end, troops would sally forth from fortified camps in order to strike rapidly at identified targets, before retiring in an effort to limit their exposure to retaliatory action. In such a manner it was intended that the insurgents would be steadily destroyed while US casualties would be kept to a minimum. But while such a strategy of raiding might make sense in a more conventional operational environment, it proved far less satisfactory in the context of an urban insurgency. The basic problem was a lack of information about exactly who the enemy was, and where they were to be found. This was exacerbated by the requirement to operate among an urban populace who were always likely to be present in considerable numbers during any given engagement. Under these conditions even the most accurate weapons were of limited value, while their use was all too certain to result in civilian casualties. Somewhat naively, US forces considered that ordinary Iraqis would accept such casualties as a necessary, if painful, step on the road to achieving their democratic freedoms.[12] But this was not the case. If anything, their effect was to alienate the people from their would-be liberators and to increase popular sympathy and support for the various groups whose declared intention was to kick them out of Iraq. This, in turn, made it easier for the insurgents to operate under the nose of their increasingly beleaguered adversaries. In short, US forces found themselves operating under a much greater burden of friction than the opposition, only to persist in acting in a manner that exacerbated their problems in this regard. In consequence they found themselves rapidly losing both soldiers and popular support.

'Culture-centric Warfare'

The emerging problems in Afghanistan and Iraq proved instrumental in inspiring a growing critique of the seemingly unalloyed faith in military technique that underpinned the Transformation process. In an influential essay on this issue, retired Major-General Robert H. Scales argued that the very technical prowess of the US armed forces had catalysed innovative corrective efforts on the part of their adversaries, and that these in turn demanded a very different, 'cultural', response from the United States. According to Scales, these adversaries had

> adapted and adopted a method of war that seeks to offset U.S. technical superiority with a countervailing method that uses guile, subterfuge, and terror mixed with patience and a willingness to die. This approach allows the weaker to take on the stronger and has proved effective against Western-style armies . . . Yet the military remains wedded to the premise that success in war is best achieved by overwhelming technological advantage. Transformation has been interpreted exclusively as a technological challenge. So far, we have spent billions to gain a few additional meters of precision, knots of speed, or bits of bandwidth. Some of that money might be better spent improving how our military thinks and studies, to create a parallel transformation based on cognition and cultural awareness.[13]

The basic notion that US forces needed to understand the societies among which they had unexpectedly found themselves operating proved influential indeed. The concept of culture – defined as that which 'provides meaning to individuals within the society' under consideration – occupied a central position in a new counter-insurgency doctrine that was already under development at the time.[14] Much like previous iterations of Western counter-insurgency doctrine dating from the 1950s and 1960s, the new US version embodied the key point that the technical challenge of destroying insurgent forces was trivial in relation to the very different kind of challenge associated with identifying and locating them in the first instance. To this end, it had to be acknowledged that the information-age systems that US forces might routinely expect to employ in order to ascertain the whereabouts of conventional targets such as tanks and aircraft were of relatively little use in the new operational environment. Finding out what was happening on the ground in down-town Baghdad required a very different approach that was better characterized as an exercise in 'human' intelligence gathering. In other words, it involved talking to the local population in order to ascertain what was afoot, because it was the locals who constituted by far the best source of intelligence in this regard, and who therefore held the key to dispelling the heavy burden of friction under which US forces had hitherto been operating. This, as a precursor, involved gaining the locals' trust, which in its turn involved demonstrating a willingness to protect them – not least by operating among them in a rather more sustained manner than had previously been the case. It was acknowledged that this would put soldiers at additional risk in the short term by increasing their exposure to enemy action. On the other hand, the previous approach of swooping down out of the blue, before shooting first and asking questions later, was clearly unsustainable over the longer term. The calculation was, therefore, that risks accepted over the short term would yield much greater benefits in terms of actionable intelligence at a later date.[15]

Raw intelligence of the type routinely collected in a counter-insurgency campaign is, however, of little use unless contextualized within a given framework of meaning, which is to say a 'culture'. This is why a great deal of emphasis was also placed on an initiative termed the 'Human Terrain System'. The basic idea underlying this initiative was to provide a means of mapping the cultural 'terrain' within which US forces found themselves operating.[16] In effect, it constituted a twenty-first-century equivalent of Britain's Ordnance Survey; for whereas the Ordnance Survey reflected the fact that the chief medium of nineteenth-century warfare was the physical geography over which armies manoeuvred, the Human Terrain System reflected the fact that contemporary warfare is fought 'amongst the people', making a capacity to map the relevant social geography highly desirable.[17] Although the system was not specifically designed to generate targeting information, it was intended (among many other things) to illuminate the likely costs and benefits flowing from the application of force in any given situation. This made it a potentially very valuable tool for counter-insurgency forces, which are routinely faced with making such difficult decisions. As General David Petraeus learned from his personal experience of operating in Iraq, such forces must always consider whether the activity they are contemplating will 'take more bad guys off the street than it creates by the way it is conducted'.[18] Thus the more one understands the social geography of the local neighbourhood, the more effectively one can judge whether shooting it up a little risks producing more benefits than it does costs in relation to one's overall political objective.

Ask questions first and shoot later
The philosophy of asking questions first and shooting later (if at all) came very much to the fore in the new US counter-insurgency doctrine. Indeed, its development benefited from the input of Petraeus who, as commander of the 101st Airborne division, had conducted his own very successful counter-insurgency campaign in northern Iraq during the early post-invasion phase. In combination with a 'surge' in troop numbers – which constituted a reversal of another key trend in the Transformation process's efficiency agenda – the generalized shift onto a counter-insurgency

footing was widely credited with helping to reverse declining US fortunes in Iraq and to stabilize the situation there.

In point of fact there is good evidence to suggest that this change in fortunes was due to local developments on the ground as much as anything that US forces contributed to the situation. By 2006 the Shia militias were evidently gaining the upper hand in the struggle with their Sunni adversaries. The latter therefore found it expedient to ally with the expanding US military, not least because of the new doctrinal emphasis it now placed on protecting the local population from attack. This in turn led to a stand off between the militias on each side of the sectarian divide, which was under-written by the US presence. By this time, too, al-Qaeda's notori-ously brutal and indiscriminate methods had worked to undermine Sunni support for its activities in Iraq. The result was that local Sunni knowledge of al-Qaeda's operatives became available to US forces, who were therefore able to direct force against them far more efficiently than had previously been the case.[19] Having said all this, however, it is also clear that the new doctrinal emphasis on protecting the Iraqi population from attack was exactly what was required in order to benefit from the changing situation on the ground, while the value of possessing local knowledge for targeting purposes was effectively demonstrated by al-Qaeda's dramatically increased vulnerability to US forces. In the event, therefore, beneficial synergies appear to have been created by the changes in local conditions and in US doctrine. At time of writing, a precarious stability – albeit punctuated by alarming outbursts of violence – has descended on Iraq, and national elec-tions have recently been held without the voting being signifi-cantly split along sectarian lines.

Back to Afghanistan

Although in many respects conditions in urban Iraq differ markedly from those in rural Afghanistan, the basic point that ISAF is fighting a war among the Afghan people remains true enough. Moreover, the steadily deteriorating security situation in

Afghanistan has been ascribed to a failure of its forces to operate in accordance with the principles of counter-insurgency that had helped improve the situation in Iraq. Shortly after assuming the position of Commander of ISAF in 2009, General Stanley McChrystal conducted his own appreciation of the challenges facing him and his forces. His conclusion was that the war remained winnable, but that this could only be achieved by means of a significant change in strategy. Hitherto, he argued, ISAF had been rather too focused on the strategic objective of destroying insurgent forces – not least by frequently resorting to air power – and that this had been working to its own ultimate detriment. The motivations behind this approach, along with the problems it had caused, were very much the same as those encountered in Iraq.

> Pre-occupied [sic] with protection of our own forces, we have operated in a manner that distances us – physically and psychologically – from the people we seek to protect. In addition, we run the risk of strategic defeat by pursuing tactical wins that cause civilian casualties or unnecessary collateral damage. The insurgents cannot defeat us militarily; but we can defeat ourselves.

The answer to this problematic state of affairs, McChrystal asserted, was to refocus efforts away from the task of attacking insurgents and onto that of protecting the Afghan people and providing for their needs 'by, with, and through' their new government. Doing so, he argued, would encourage the people to reciprocate in kind, by supporting the government and cooperating in defeating the Taliban. McChrystal recognized, however, that co-opting the Afghan people in such a manner would involve working in a 'complex social landscape' that required to be understood far more thoroughly than had previously been the case.

> This complex environment is challenging to understand, particularly for foreigners. For this strategy to succeed, ISAF

leaders must redouble efforts to understand the social and political dynamics of areas [sic] all regions of the country and take action that meets the needs of the people.[20]

In all of this McChrystal would seem to be correct. Any hope of creating a minimum acceptable political outcome in Afghanistan (which is to say a state willing and able to deny al-Qaeda sanctuary) is not going to lie in the comprehensive destruction of the Taliban via the application of highly accurate firepower. To be sure, armed force will have a part to play, but rather than constituting the main effort it ought to be a relatively small and circumscribed one. Force will, in other words, need to be applied in a manner that both reflects, and helps shape, the political context, rather than one that altogether ignores it – as the strategic ideas underpinning the Transformation initiative were all too prone to do. At time of writing, it remains to be seen whether this can be achieved, although there are some hopeful signs in this regard.

Most recently, ISAF's offensive in southern Iraq – operation *Moshtarak* – has demonstrated a keen awareness of the problems associated with using armed force in a counter-insurgency context. In a significant departure from traditional practice, the operation was announced ahead of time, thereby sacrificing the element of surprise that is highly prized in other strategic contexts. The purposes of doing so were two-fold. First, Afghan civilians were thereby provided with an opportunity to leave the area so that they would not become casualties of the offensive. Secondly, a degree of forewarning also provided elements of the Taliban, who might be co-opted into some form of political settlement with the Afghan government, with an incentive to negotiate. The potential political advantages flowing from these arrangements were judged to outweigh the military risks associated with allowing the Taliban an opportunity to bolster its defences. In the same spirit, ISAF deliberately reduced its reliance on air power and other weapons of the type that were prone to inflicting undue levels of collateral damage. Another important innovation came in the form of a clear commitment to remain among the Afghan people who were liberated from Taliban control, in order to protect both them and the construction work that was scheduled to begin in the immediate wake of the offensive. This was rendered more feasible by the fact that large numbers of Afghan security forces were now available for employment in both combat and security roles.[21]

Technique and time

The course of the wars in Iraq and Afghanistan once again demonstrates that military technique can provide no certain substitute for a clear understanding of the political context in which one's armed forces are expected to operate. To be sure, the initial phases of both conflicts saw the US military prevail in the face of opposition that was ill prepared to conduct a toe-to-toe fight. On the other hand, Washington's failure to secure a stable political settlement from its early military successes provided its adversaries with a vital opportunity to adopt new techniques for prolonging the struggle and to hit back, inflicting far more casualties than they had previously managed. In its turn, the challenge posed by terrorism and insurgency demanded a countervailing response from the US military, which emerged in the form of a new counter-insurgency doctrine supported by a Human Terrain System designed to permit more refined cost-benefit calculations in relation to the use of force in any given socio-political context.

These innovations have themselves taken time to implement, however, during which casualties have continued to mount and new challenges have also come to the fore. The most pressing of these is the increased use of improvised explosive devices (IEDs) by the enemy. A conspicuous feature of the war in Iraq, they have clearly become the Taliban's weapon of choice in Afghanistan where to date they have proved highly efficient killers of ISAF troops. Improvised they may be, but they are also 'high-tech' in the sense of being remarkably fit for purpose. Different basic designs have been optimized to attack a range of targets, including armoured vehicles. Moreover a great many of them contain very little (and sometimes no) metal content, rendering them very difficult to detect. Thus, despite creating an organization with a multi-billion dollar budget that is dedicated to tackling the IED threat, the United States is by no means currently on top of this latest challenge. In 2009 IEDs were responsible for 75 per cent of the casualties suffered by ISAF.[22] Indeed in this particular aspect of warfare, the Taliban would seem to enjoy a pronounced technical advantage over their adversaries. This is not to say that the IED challenge cannot, in principle, be overcome; but it does seem reasonable to question whether it can be overcome sufficiently rapidly for ISAF casualties to be kept within politically manageable bounds. Having initially failed to deny its adversaries the time necessary to respond to its superior military technique, the United States has now found it being played at its own game. And that this is so is due to the fact that Washington did not understand the character of the political forces it would generate by invading Iraq and Afghanistan.

Afterword

Iraq and Afghanistan have hitherto been considered the main 'fronts' (to use a somewhat anachronistic term) in the War on Terror. The problem is, however, more widespread than this. Islamist influences are by no means limited to those areas on which the West has chosen to focus its military power. Left unchecked such influences have the potential to grow in scale and scope, threatening allies and interests in the Middle East and other regions, along with the very security of the West itself. Whatever happens in Afghanistan, therefore, some form of global engagement is required in order to forestall threatening developments elsewhere. On the other hand, there no longer exists any appetite for the kind of adventures in regime change that so enticed the Bush Jr administration. Other peoples' politics turned out to be rather more complicated and unpredictable than the Neoconservative world-view cared to admit, and it also emerged that no amount of military-technical superiority was able to remedy this disappointing discovery. Having inherited a disastrous legacy of political and technical hubris from its predecessor, the Obama administration has therefore found itself facing the perennial challenge associated with strategic decision-making: the need to establish a more realistic balance between the risks of not doing enough to ward off emerging threats, and the risks of doing too much – of taking inflammatory and expensive initiatives that ultimately create more problems than they solve.

Clearly, advanced military technique will have an important role to play in any strategy that is designed to navigate between the two extremes of leaving threats unchecked and engaging in further bouts of regime change. Stand-off weaponry, employed perhaps in combination with special forces, represents one obvious possibility in this regard. As operations in Pakistan suggest, the capacity to attack and destroy the key personnel and assets of terrorist organizations will help to disrupt their operations and keep them on the back foot. On the other hand, there will be severe limits imposed on the extent to which the United States can wage a war of long-range bombardment along these lines. It must be expected that any adversary that is threatened in such a manner will sooner or later introduce technical counter measures by way of response. More importantly, however, finding such targets without recourse to local knowledge promises to be well-nigh impossible, while anything in the way of speculative bombardment is all too likely to cause collateral damage that an enemy can twist to his own advantage. This means that any such operations will require cooperation from the states within whose borders they occur. Without this, they will be ineffective, if not counter-productive.

And if co-operation of this type will be a necessary precursor to effective military action, it will also very likely come with political strings attached. The 'with us or against us' mentality of the Bush administration simply cannot do justice to the highly nuanced character of politics in areas such as the Middle East, where a great deal habitually rests on avoiding stark choices of this kind. This will, in turn, demand an astute appreciation of the local political contexts – the 'human terrain', so to speak – within which co-operating states are located, of the leeway they enjoy for accommodating Western agendas when their own populations may very well entertain profound misgivings about these. Only by understanding the limits beyond which the application of force will quickly become politically counter-productive will it be possible to shape the formulation of strategic objectives accordingly. The result will almost certainly be to rule out ambitious acts of surgical destruction, of the kind envisaged by the most enthusiastic proponents of the Transformation initiative, in favour of more modest efforts intended to manage threats over the longer term. In this regard the military dimension of the War on Terror will need to acquire a more limited, protective character until other developments in the area of 'soft' power (which lie beyond the scope of our present concerns) have a chance to gain purchase and bring a more definitive end to the threat posed by Islamist terrorism.

At a fundamental level, therefore, the challenge associated with the use of force in today's War on Terror is no different from that faced by other strategic actors in the past. The challenge remains one of striking an appropriate balance between attempting too little and too much. Success in this regard demands the application of judgement, and the more that our judgement is informed by an accurate understanding of our adversaries, the sounder it will be. However much we might like warfare to be reducible to a purely technical exercise of the type that plays best to Western strengths, it remains first and foremost a continuation of politics. We reject Clausewitz's time-honoured message at our peril.

Notes

Introduction

1   In regard to Iraq the term *regime change* would seem to be something of a misnomer, given that attention was focused simply on toppling the existing regime as opposed to replacing it with another.

2   Hans Morgenthau, 'We Are Deluding Ourselves in Vietnam', *New York Times Magazine*, 18 April 1965.

3   For an excellent discussion of the many obstacles to effective strategy see Richard K. Betts, 'Is Strategy an Illusion?', *International Security* 25 (2000): 5–50.

4   Peter Paret, 'Introduction', in *The Makers of Modern Strategy: from Machiavelli to the Nuclear Age*, ed. Peter Paret (Oxford: Clarendon, 1986), 3.

5   Thomas C. Schelling, *The Strategy of Conflict* (Cambridge, Mass.: Harvard University Press, 1960), 86.

6   An alternative is to strike punitively at a selection of non-military assets that an adversary is understood to value, but that do not have any bearing on his ability to continue fighting per se.

7   For the key features of his argument on this point see Carl von Clausewitz, *On War*, trans. Colonel J.J. Graham (New York: Barnes & Noble, 2004), 1–19.

8   For a different, but related, discussion of why strategy is difficult see Colin Gray, 'Why Strategy Is Difficult', *Joint Force Quarterly* 22 (1999): 6–12.

9   Clausewitz, *On War*, 649.

10   Isaiah Berlin, *The Sense of Reality: Studies in Ideas and their History*, ed. Henry Hardy (London: Chatto & Windus, 1996), 46.

11   Clausewitz, *On War*, 129.

12   Gray, 'Why Strategy Is Difficult', 10.

13   My approach echoes that of Quincy Wright, *A Study of War*, 2nd edn (Chicago, Ill.: University of Chicago Press, 1965), 291–2, for whom the 'technique of war concerns, on the one hand, the instruments (weapons and organizations) with which war is carried on, on the other hand, the utilization of these instruments (operations and policies) to achieve the objects of war.'

14   Baron Henri de Jomini, *Précis de l'Art de la Guerre* '(Paris: n.p., 1838). According to our terminology, therefore, Jomini's text is a *technology*, which is to say a study of technique.

15   Clausewitz, *On War*, 1–19 with additional comments on 52–62.

16   Thomas C. Schelling, *Arms and Influence* (New Haven, Conn.: Yale University Press, 1966), 19–20.

17   Edward N. Luttwak, *Strategy: The Logic of War and Peace* (Cambridge, Mass.: Belknap, 1987), 27–31.

18   Clausewitz, *On War*, 1, 128.

Chapter 1

1   Charles Ingrao, 'Paul W. Schroeder's Balance of Power: Stability or Anarchy?', *International History Review* 16 (1994): 686–7.

2   Christopher Duffy, *The Military Experience in the Age of Reason* (London: Routledge & Keegan Paul, 1987), 11.

3   Marshal Maurice de Saxe, 'My Reveries Upon the Art of War', in *Roots of Strategy: A Collection of Military Classics*, trans. and ed. Brig. Gen. Thomas R. Phillips (Mechanicsburg, Pa.: Stackpole, 1985), 298–9.

4   Frederick the Great, *Instructions for his Generals*, trans. Brigadier General Thomas R. Phillips (Harrisburg, Pa.: Military Service Publishing Company, 1944), 95.

5   Colonel J.F.C. Fuller, *The Foundations of the Science of War* (London: Hutchinson, 1926), 19.

6   Emmanuel Joseph Sieyès, *Qu'est-ce que le Tiers État?* (Paris: Champs Flammarion, 1988), 127 with original emphasis.

7   Edmund Burke, *The Writings and Speeches of Edmund Burke, Vol. IX: The Revolutionary War, 1794–1797*, ed. R.B. McDowall (Oxford: Clarendon, 1991), 267, 290 with original emphasis.

8   R.R. Palmer, 'Frederick the Great, Guibert, Bülow: From Dynastic to National War', in *Makers of Modern Strategy: from Machiavelli to the Nuclear Age*, ed. Peter Paret (Oxford: Clarendon, 1986), 100, n. 15. Such a comparison probably underplays the transformative role played by the French Revolution on the character of strategy, given that Frederick himself fought an unusually large number of battles by the standards of his day.

9   David A. Bell, *The First Total War: Napoleon's Europe and the Birth of Modern Warfare* (London: Bloomsbury, 2007), 135–6.

10   Lazare Carnot, 'Système générale des operations militaries de la campagne prochaine', in *Correspondence générale de Carnot, Vol. IV: novembre 1793 – mars 1795*, ed. E. Charavay (Paris: Imprimerie Nationale, 1907), 283.

11   Gunther Rothenburg, *The Art of Warfare in the Age of Napoleon* (London: Batsford, 1977), 36.

12   Baron Henri de Jomini, *Précis de l'Art de la Guerre* (Paris: n.p., 1838), 201–2.

13   In point of fact he was almost beaten at Marengo, and was saved only by the timely intervention of General Louis Desaix. But Desaix was himself

killed during the battle, providing Bonaparte with an opportunity to pinch all the glory.

14 M. de Bourrienne, *Memoirs of Napoleon Bonapart*, Vol. I (London: Richard Bentley, 1836), 118.

15 Cited in Jean Tulard, *Napoleon: The Myth of the Saviour*, trans. Teresa Waugh (London: Weidenfeld Nicolson, 1984), 307.

16 David Gates, *The Spanish Ulcer: A History of the Peninsular War* (London: George Allen & Unwin, 1986), 468–9.

17 Minard's *Carte Figurative des pertes successives en hommes de l'Armée Francaise dans la campagne de Russie 1812–1813* is reproduced in Edward R. Tufte, *The Visual Display of Quantitative Information*, 2nd edn (Cheshire, Conn.: Graphics Press, 2001), 41.

18 Jomini, *Précis de l'Art de la Guerre*, 58.

19 Charles J. Esdaile, *The Wars of Napoleon* (London: Longman, 1995), 29.

Chapter 2

1 Otto Pflanze, *Bismarck and the Development of Germany: The Period of Unification, 1815–1871* (Princeton, N.J.: Princeton University Press, 1963), 10–11.

2 Otto von Bismarck, *The Love Letters of Bismarck: Being Letters to His Fiancée and Wife, 1846–1889, Authorized by Prince Herbert von Bismarck and Translated from the German under the Supervision of Charlton T. Lewis* (New York: Harper Brothers, 1901), 410. Bismarck's wife appears to have been kept rather well informed on Prussian matters of state, although sadly deprived of the French crinoline (p. 378) she evidently desired.

3 Helmuth von Moltke, *Moltke on the Art of War: Selected Writings*, trans. and ed. Daniel J. Hughes and Harry Bell (Novato, Calif.: Presidio, 1993), 127.

4 In the French case this included the time required to set up the new factory although some orders were also placed abroad. William H. McNeill, *The Pursuit of Power: Technology, Armed Force, and Society since A.D. 1000* (Oxford: Blackwell, 1983), 235–6.

5 Carl von Clausewitz, *On War*, trans. Colonel J.J. Graham (New York: Barnes & Noble, 2004), 2.

6 Moltke, *Art of War*, 36, 44–5, 176.

7 Ibid., 219.

8 Helmuth von Moltke, *Moltke's Military Correspondence 1870–71*, ed. Spencer Wilkinson (Aldershot: Gregg Revivals, 1991), 28.

9 Moltke, *Art of War*, 46.

10 Cited in Pflanze, *Bismarck and the Development of Germany*, 458, n. 1.

11 Otto Prince von Bismarck, *Bismarck: The Man and the Statesman*, trans. A.J. Butler (London: Smith, Elder & Co., 1898), 105.

12 Ibid., 106.

13 Ibid., 41.

14 Ibid., 107–8.

15 Cited in George Eliot Buckle (in succession to W.F. Monypenny), *The Life of Benjamin Disraeli, Earl of Beaconsfield, Volume V: 1868–1876* (London: John Murray, 1920), 133.

16 Helmuth von Moltke, *Essays, Speeches, and Memoirs of Field-Marshal Count Helmuth von Moltke*, Volume II, trans. Charles Flint McClumpha, Major C. Barter and Mary Herms (New York: Harper & Brothers, 1893), 127.

17 Colonel G.F.R. Henderson, *The Science of War: A Collection of Essays and Lectures 1892–1903*, ed. Captain Neill Malcolm (London: Longmans, Green & Co., 1905), 9.

18 Moltke, *Essays, Speeches, and Memoirs of Field-Marshal Count Helmuth von Moltke*, 137.

19 Gerhard Ritter, *The Schlieffen Plan: Critique of a Myth*, trans. Andrew and Eva Wilson (London: Oswald Wolf, 1958), 17–21.

20 Munroe Smith, 'Military Strategy Versus Diplomacy in Bismarck's Time and Afterward', *Political Science Quarterly* 30 (1915): 69.

21 During the 1880s Tsar Alexander regarded Wilhelm as nothing more than a 'badly raised boy' and maintained his distance accordingly. For his part Wilhelm considered Alexander's lacklustre son, Nicholas, 'only fit to live in a country house and grow turnips', and practised on him to some effect once he had succeeded his father. The least that can be said for Nicholas II, however, was that he ultimately acquiesced in the advice of his ministers in the matter of relations with Germany. Barbara Tuchman, *The Guns of August* (New York: Macmillan, 1962), 22–4.

22 Hajo Holborn, 'Moltke and Schlieffen: The Prussian-German School', in *Makers of Modern Strategy: Military Thought from Machiavelli to Hitler*, ed. Edward Meade Earle (Princeton, N.J.: Princeton University Press, 1943), 190.

23 Prince von Bülow, *Memoirs, Vol. II: 1903–1909*, trans. F.A. Voigt (London: Putnam, 1931), 73, 'strategical considerations' in our context meaning military-technical considerations.

24 Ibid., 73–4. To be sure, Bülow almost certainly recounted events in such a manner as to paint himself in the best possible light. Nevertheless, some version of the meeting between Wilhelm and Leopold did take place, and if Bülow chose to represent the emperor in such an unsettling light it was because he knew that his account would be considered plausible by those familiar with Wilhelm's ways. For further discussion of Bülow and related matters see: Ritter, *Schlieffen Plan*, 78–96; L.C.F. Turner, 'The Significance of the Schlieffen Plan', in *The War Plans of the Great Powers, 1880–1914*, ed. Paul Kennedy (London: George Allen & Unwin, 1979), 205–7.

25 Alfred von Schlieffen, *Alfred von Schlieffen's Military Writings*, trans. and ed. Robert T. Foley (London: Frank Cass, 2003), 198–9.

26 Gunther E. Rothenburg, 'Moltke, Schlieffen, and the Doctrine of Strategic Envelopment', in *The Makers of Modern Strategy: from Machiavelli to the Nuclear Age*, ed. Peter Paret (Oxford: Clarendon, 1986), 314.

27 Ritter, *Schlieffen Plan*, 66.

28 Alistair Horne, *The Price of Glory: Verdun 1916* (London: Macmillan, 1962), 24.

29 André Corvisier, ed., *A Dictionary of Military History*, trans. Chris Turner, rev. John Childs (Oxford: Basil Blackwell, 1994), 470.

30 Holborn, 'Moltke and Schlieffen', 204.

31 Isaiah Berlin, *The Sense of Reality: Studies in Ideas and their History*, ed. Henry Hardy (London: Chatto & Windus, 1996), 49.

32 Michael Howard, *The Franco-Prussian War* (London: Rupert Hart-Davis, 1961), 454.

Chapter 3

1 I.S. Bloch, *Is War Now Impossible? Being an Abridgement of the War of the Future in its Technical, Economic and Political Relations* (Aldershot: Gregg Revivals, in association with the Department of War Studies King's College London, 1991), xi, xli.

2 Beatrice Heuser, *Reading Clausewitz* (London: Pimlico, 2002), 119; General Ludendorff, *The Nation at War*, trans. A.S. Rappoport (London: Hutchinson, n.d. [1936]).

3 Although in this latter instance he did no more than execute a plan that had already been drawn up by the senior staff officer on the scene, Colonel Max Hoffmann.

4 Ludendorff, *The Nation at War*, 23-4.

5 Hans Speier, 'Ludendorff: The German Concept of Total War', in *Makers of Modern Strategy: Military Thought from Machiavelli to Hitler*, ed. Edward Meade Earle (Princeton: N.J.: Princeton University Press, 1943), 308.

6 Liddell Hart, *Europe in Arms* (London: Faber & Faber, 1937), 221.

7 'Pacifism and the War: A Controversy. By D.S. Savage, George Woodcock, Alex Comfort, George Orwell', *Partisan Review* 9 (1942), 414 with original emphasis.

8 Savage, however, would presumably have claimed that the latter fate was the lesser of two evils. For a first-hand account of his philosophy, and the sacrifices entailed by living it, see his 'Testament of a Conscientious Objector', in *The Objectors*, ed. Clifford Simmons (Isle of Man: Anthony Gibbs & Phillips, n.d. [1965]), 82–122.

9 Colonel J.F.C. Fuller, *The Reformation of War* (London: Hutchinson, 1923); Captain B.H. Liddell Hart, *Paris, or the Future of War* (London: Kegan, Paul, Trench, Trubner, 1925); Giulio Douhet, *Command of the Air*, trans. Dino Ferrari (London: Faber & Faber, 1943); General De Gaulle, *The Army of the Future* (London: Hutchinson, n.d. [1941]); Major-General Heinz

Guderian, *Achtung-Panzer! The Development of Armoured Forces, their Tactics and Operational Potential*, trans. Christopher Duffy (London: Arms & Armour Press, 1992).

10   Colonel J.F.C. Fuller, *On Future Warfare* (London: Sifton Praed, 1928), 153.

11   Ludendorff, *The Nation at War*, 94–5.

12   Victor Wallace Germains, *The 'Mechanization' of War* (London: Sifton Praed, 1927), 117, 127–8, 183.

13   Guderian, *Achtung-Panzer!*, 205–6.

14   A good example of the speed with which relevant responses to German technique were formulated is provided by F.O. Miksche's *Blitzkrieg* (London: Faber & Faber, 1941), which Tom Wintringham described in his introduction (p. 13) as 'a textbook of modern warfare.'

15   Major-General J.F.C. Fuller: *Lectures on F.S.R. III (Operations Between Mechanized Forces)*, (London: Sifton Praed, 1932), esp. 129–33; *Towards Armageddon: The Defence Problem and its Solution* (London: Lovat Dickson, 1937), 239.

16   Liddell Hart, *The Defence of Britain* (London: Faber & Faber, 1939), 120–1.

17   Ibid., 107.

18   Ibid., 121, 124.

19   Ibid., 206–8, although in practice the Maginot line was not meant as an alternative to major offensive operations. Rather, it was designed to avoid the necessity of extensive peacetime preparations for war by providing a defensive glacis behind which France could mobilize if attacked. The expectation was that Franco-British forces would move onto the offensive once they were thoroughly prepared.

20   Liddell Hart, *When Britain Goes to War: Adaptability and Mobility* (London: Faber & Faber, 1935), 49–50, 55–6.

21   Liddell Hart, *Europe in Arms*, 212.

22   In the wake of the 1938 Munich crisis, Chamberlain was perturbed to learn from British intelligence that Hitler referred to him as an 'arsehole' and derided his habit of carrying an umbrella. It is possible that an MI5 informant was exaggerating for effect here. Nevertheless, the resulting picture of Hitler was rather more accurate than that possessed by Chamberlain. Christopher Andrew, *The Defence of the Realm: The Authorized History of MI5* (London: Penguin, 2010), 205–6, 909 n. 98.

23   George Orwell, *The Collected Essays, Journalism and Letters of George Orwell, Volume II: My Country Right or Left, 1940–1943*, ed. Sonia Orwell and Ian Angus (London: Secker & Warburg, 1968), 13–14, 248.

24   When those aircraft that did not attack their targets were included, the proportion getting within five miles fell to just one in five. Charles Webster and Noble Frankland, *The Strategic Air Offensive Against Germany 1939–1945, Volume IV: Annexes and Appendices* (London: HMSO, 1961), 205.

25   Albert Speer, *Inside the Third Reich: Memoirs by Albert Speer*, trans. Richard and Clara Winston (London: Weidenfeld & Nicolson, 1970), 284.

26   Charles Webster and Noble Frankland, *The Strategic Air Offensive Against Germany 1939–1945, Volume II: Endeavour* (London: HMSO, 1961), 155, 194.

27   For an interesting account see Richard Simpkin, in association with John Erickson, *Deep Battle: The Brainchild of Marshall Tukhachevskii* (London: Brassey's, 1987).

28   Cited in John Erickson, *The Road to Stalingrad: Stalin's War With Germany, Volume 1* (London: Weidenfeld & Nicolson, 1975), 5.

29   B.H. Liddell Hart, *The Revolution in Warfare* (London: Faber & Faber, 1946), 85–9, with original emphasis.

Chapter 4

1   Franklin D. Roosevelt, 'Annual Message to Congress on the State of the Union', 6 January 1941, retrieved from John T. Woolley and Gerhard Peters, *The American Presidency Project* [online], Santa Barabara, Calif., http://www.presidency.ucsb.edu/ws/index.php?pid=16092, accessed April 2010.

2   General George C. Marshall, *The Winning of the War in Europe and the Pacific: Biennial Report of the Chief of Staff of the United States Army 1943 to 1945, to the Secretary of War* (New York: Simon & Schuster, 1945), 117.

3   Alexis de Tocqeville, *De la démocratie en Amérique* (Paris: Garnier-Flammarion, 1981), 329.

4   Had matters been decided by a single clash, the total number of dead and wounded would have come to around 3,575, which was the casualty bill for the war's initial pitched battle, First Bull Run (1861). James M. McPherson, *Battle Cry of Freedom: The American Civil War* (Oxford: Oxford University Press, 1988), 347.

5   Although Henderson preferred to explain the matter in terms of 'racial instinct' rather than political philosophy. Colonel G.F.R. Henderson, *The Science of War: A Collection of Essays and Lectures 1892–1903*, ed. Captain Neill Malcolm (London: Longmans, Green & Co., 1905), 129.

6   John J. Pershing, *My Experiences in the World War* (London: Hodder & Stoughton, 1931), 23–4, 142–4.

7   Nor did the Army's ground forces receive a great many of the most intelligent men, as these were required for technical duties in the air forces. Alan R. Millet, 'The United States Armed Forces in the Second World War', in *Military Effectiveness, Volume III: The Second World War*, ed. Allen R. Millet and Williamson Murray (Boston, Mass.: Allen & Unwin, 1988), 60.

8   Marshall, *The Winning of the War*, 99, 101–2, 119.

9   William Mitchell, *Winged Defense: The Development and Possibilities of Modern Airpower – Economic and Military* (New York: G.P. Putnam's Sons,

1925); Alexander Seversky, *Victory Through Air Power* (New York: Simon & Schuster, 1942).

10   FM 17–100, *Armored Command Field Manual: The Armored Division* (Washington, DC: US Government Printing Office, 1944), 2

11   In a raid mounted on 17 August, 36 out of the 230 aircraft committed were declared missing, for a loss rate of 15.7 per cent. Moreover some aircraft that made it back to base never flew again. Martin Middlebrook, *The Schweinfurt-Regensburg Mission: American Raids on 17 August 1943* (London: Cassell, 2000), 280.

12   Albert Speer, *Inside the Third Reich*, trans. Richard and Clara Winston (London: Weidenfeld & Nicolson, 1970), 285.

13   Marshall, *The Winning of the War*, 107–8.

14   This figure includes prisoners of war. John Erickson, *The Road to Berlin: Stalin's War with Germany, Volume 2* (London: Weidenfeld & Nicolson, 1983), ix.

15   Richard Overy, *Russia's War* (London: Allen Lane The Penguin Press, 1998), 288.

16   Marshall, *The Winning of the War*, 107.

17   X [George Kennan], 'The Sources of Soviet Conduct', *Foreign Affairs* 25 (1946–47): 575.

18   NSC 20/1, 'US Objectives with Respect to Russia', 18 August 1948, reprinted in *Containment: Documents on American Policy and Strategy, 1945–1950*, ed. Thomas H. Etzold and John Lewis Gaddis (New York: Columbia University Press, 1978), 173–203. Clausewitz is cited on p. 174.

19   NSC-68, 'United States Objectives and Programs for National Security', 14 April 1950, reprinted in *Containment*, ed. Etzold and Gaddis, 385–442.

20   United Nations General Assembly, A/1435, 'The Problem of the Independence of Korea', 7 October 1950, 2.

21   Reprinted in Douglas MacArthur, *Reminiscences* (London: Heinemann, 1964), 387–8.

22   Ridgway would later write an account of the war, whose subtitle suggests he understood very clearly the new challenges associated with the application of force. Matthew B. Ridgway, *The War in Korea: How We Met the Challenge, How All-Out Asian War Was Averted, Why MacArthur Was Dismissed, Why Today's War Objectives Must Be Limited* (London: Barrie & Rockliff, The Crescent Press, 1968).

23   William W. Kaufmann, 'Limited Warfare', in *Military Policy and National Security*, ed. William W. Kaufmann (Princeton, N.J.: Princeton University Press, 1956), 116–17.

24   NSC 20/4, 'US Objectives with Respect to the USSR to Counter Soviet Threats to US Security', 23 November 1948, reprinted in *Containment*, ed. Etzold and Gaddis, 209–10.

25 Marshall, *The Winning of the War*, 117–23; Aaron L. Freidberg, 'Why Didn't the United States Become a Garrison State?', *International Security* 16 (1992): 125–8.

26 Harold D. Lasswell, 'Sino-Japanese Crisis: The Garrison State versus the Civilian State', *China Quarterly* (Shanghai) 2 (1937): 643.

27 NSC-162/2, 'A Report to the National Security Council by the Executive Secretary on Basic National Security Policy', 30 October 1953, 18.

28 John Foster Dulles, 'The Evolution of Foreign Policy', *Department of State Bulletin*, 25 January 1954: 107–8.

29 William W. Kaufmann, 'The Requirements of Deterrence', in *Military Policy and National Security*, ed. Kaufmann, 23.

30 Henry A. Kissinger, *Nuclear Weapons and Foreign Policy* (New York: Harper & Brothers, for the Council on Foreign Relations, 1957), 87.

Chapter 5

1 Bernard Brodie, *The Absolute Weapon: Atomic Power and World Order* (New York: Harcourt Brace, 1946), 76.

2 Dwight D. Eisenhower, 'The President's News Conference', 12 January 1955, retrieved from John T. Woolley and Gerhard Peters, *The American Presidency Project* [online], Santa Barabara, Calif., http://www.presidency.ucsb.edu/ws/index.php?pid=10232, accessed May 2010.

3 Eisenhower had made a study of Clausewitz during his early military career, and some of it appears to have stuck. See Christopher Bassford, *Clausewitz in English: The Reception of Clausewitz in Britain and America 1815–1945* (Oxford: Oxford University Press, 1994), 157–62.

4 Albert Wohlstetter, 'The Delicate Balance of Terror', Rand Report P-1472 (Santa Monica, Calif.: Rand, 1958); James E. King Jr, 'Nuclear Plenty and Limited War', *Foreign Affairs* 35 (1957): 238–56.

5 This helps to explain Eisenhower's famous predilection for golf: being out on the course hindered the efforts of politically proactive spirits, such as Dulles, to reach him. Many saw the golf as symptomatic of nothing more than presidential laziness, but an extensive series of interviews with Eisenhower subsequently led Walter Cronkite to conclude that the former president had possessed a strong grasp of the issues that had faced him during his time in office. Walter Cronkite, *A Reporter's Life* (New York: Knopf, 1996), 229, 236–7.

6 This is the compelling thesis of Campbell Craig, *Destroying the Village: Eisenhower and Thermonuclear War* (New York: Columbia University Press, 1998).

7 McNamara's ideas on the acquisition and employment of armed force receive detailed treatment in: William W. Kaufmann, *The McNamara Strategy*

(New York: Harper & Row, 1964); Alain C. Enthoven and Wayne K. Smith, *How Much is Enough? Shaping the Defense Program 1961–1969* (New York: Harper & Row, 1971).

8   John Fitzgerald Kennedy, *The Strategy of Peace*, ed. Allan Nevins (New York: Harper and Brothers, 1960), 37–8, cited in Kaufman, *The McNamara Strategy*, 41.

9   General Maxwell D. Taylor, *The Uncertain Trumpet* (New York: Harper, 1959), 6.

10   William P. Mako, *U.S. Ground Forces and the Defense of Central Europe* (Washington, DC: Brookings, 1983), 17.

11   By way of contrast, 'countervalue' targeting was code for bombing Soviet cities.

12   Strategic Air Command was the branch of the US Air Force tasked with delivering nuclear weapons by long-range aircraft and missiles.

13   The story of RAND and its relationship with the Counterforce strategy has received a lively treatment by Fred Kaplan, *The Wizards of Armageddon* (New York: Simon & Schuster, 1983).

14   Herman Kahn, *On Escalation: Metaphors and Scenarios* (London: Pall Mall Press, 1965).

15   'Remarks by Secretary McNamara, NATO Ministerial Meeting, 5 May 1962, Restricted Session' (Top Secret), p. 2.

16   Hans J. Morgenthau, 'The Four Paradoxes of Nuclear Strategy', *The American Political Science Review* 58 (1964): 28.

17   T.C. Schelling, 'Controlled Response and Strategic Warfare', *Adelphi Paper 19* (London: IISS, 1965), 6.

18   Colonel General P. Ivashutin, tr. Svetlana Savranskaya, 'Strategic Operations of the Nuclear Forces', 28 August 1964, http://php.isn.ethz.ch/collections/colltopic.cfm?id=16248&lng=en, accessed May 2010.

19   John F. Kennedy, 'Radio and Television Report to the American People on the Soviet Arms Buildup in Cuba', 22 October 1962, retrieved from the John F. Kennedy Presidential Library and Museum, http://www.jfklibrary.org/Historical+Resources/Archives/Reference+Desk/Speeches/JFK/003POF03CubaCrisis10221962.htm, accessed May 2010.

20   A private undertaking was also made to withdraw US missiles from Turkey, thereby further sweetening the pill for the Kremlin. They were, in any case, scheduled to be replaced by 'Polaris' submarine-launched missiles.

21   Enthoven and Smith, *How Much is Enough?*, 175, 207–8.

22   A MIRV system was first deployed on the 'Minuteman III' missile, and could accurately deliver three 170-kiloton warheads onto separate targets. By way of contrast, the original 'Minuteman' delivered a single warhead with a yield somewhat in excess of one megaton.

23   Richard Nixon, 'Third Annual Report to the Congress on Foreign Policy', 9 February 1972, retrieved from John T. Woolley and Gerhard Peters,

The American Presidency Project [online], Santa Barabara, Calif., http://www.presidency.ucsb.edu/ws/index.php?pid=3736&st=foreign&st1=policy, accessed May 2010.

24   'National Security Decision Memorandum 242', 17 January 1974 (Top Secret / Sensitive), p. 2.

25   Presidential Directive / NSC-59, 'Nuclear Weapons Employment Policy', 25 July 1980 (Secret with Top Secret attachment), unpaginated.

26   Leaked defence planning guidance cited by Richard Halloran, 'Pentagon Draws up First Strategy for Fighting a Long Nuclear War', *New York Times*, 30 May 1982.

27   Or perhaps 'Rube Goldberg' is more appropriate in this particular context.

28   NSDD 12, 'Strategic Forces Modernization Program', 1 October 1981 (partial unclassified version), pp. 1–2.

29   Bernard Brodie, 'The Development of Nuclear Strategy', *ACIS Working Paper No. 11* (Los Angeles, Calif.: Center for Arms Control and International Security, 1978), 24.

30   Thomas C. Schelling, *The Strategy of Conflict* (Cambridge, Mass.: Harvard University Press, 1960), 187–203.

31   Lawrence Freedman, *The Evolution of Nuclear Strategy*, 2nd edn. (London: Macmillan in association with the International Institute for Strategic Studies, 1989), 433.

Chapter 6

1   Phrases such as 'graduated military pressure' abound in Volume III of the *Pentagon Papers: The Defense Department History of United States Decision-making on Vietnam*, Gravel Edn (Boston, Mass.: Beacon Press, 1971–72), which can be retrieved via http://www.mtholyoke.edu/acad/intrel/pentagon/pent1.html, accessed June 2010.

2   This is a message that is constantly repeated in Vo Nguyen Giap, *The Military Art of People's War: Selected Writings of Vo Nguyen Giap*, ed. Russell Slater (New York: Monthly Review Press, 1970).

3   For his generally optimistic account of operations in Vietnam, see General W.C. Westmoreland, USA, Commander, U.S. Military Assistance Command, Vietnam *Report on Operations in South Vietnam January 1964 – June 1968* (Washington, DC: US Government Printing Office, n.d. [1968]).

4   Although as Walter Cronkite, *A Reporter's Life* (New York: Knopf, 1996), 258, 260–1, observes, his own role in the story is contested.

5   Robert E. Osgood, *Limited War Revisited* (Boulder, Col.: Westview Press, 1979), 44–5.

6   Robert S. McNamara, with Brian VanDeMark, *In Retrospect: The Tragedy and Lessons of Vietnam* (New York: Times Books, 1995), 32, 322.

7 General William C. Westmoreland, *A Soldier Reports* (New York: Double-day, 1976), 142–3.

8 This sentiment ultimately achieved official expression in the so-called 'Weinberger Doctrine', which flatly rejected the graduated use of force. 'The Uses of Military Power', Remarks Prepared for Delivery by the Hon. Caspar W. Weinberger, Secretary of Defense, to the National Press Club, Washington, D.C., 28 November 1984, http://www.pbs.org/wgbh/pages/frontline/shows/military/force/weinberger.html, accessed June 2010.

9 This at least was how the army viewed matters. See FM 100–5 *Operations* (Washington, DC: Headquarters, Department of the Army, 1976), 1–1, with original emphasis.

10 The accuracy of these PGMs resulted from their ability to home onto their targets in response to 'information' provided by their operators.

11 For a balanced collection of essays on these matters see Geoffrey Kemp, Robert L. Pfaltzgraff, Jr & Uri Ra'anan, *The Other Arms Race: New Technologies and Non-Nuclear Conflict* (Lexington, Mass.: Lexington Books, 1975).

12 For discussion of these matters with particular reference to the tank, see John Stone, *The Tank Debate: Armour and the Anglo-American Military Tradition* (Amsterdam: Harwood, 2000), 76–9.

13 Influential in this regard was an essay by Edward N. Luttwak, 'The Operational Level of War', *International Security* 5 (1980–81): 61–79.

14 Ben Dankbaar, 'Alternative Defense Policies and the Peace Movement', *Journal of Peace Research* 21 (1984): 145.

15 FM 100–5 *Operations* (Washington, DC: Department of the Army, 1986), 1.

16 Colin Powell, with Joseph E. Persico, *A Soldier's Way: An Autobiography* (London: Hutchinson, 1995), 509–10.

17 Anne Leland and Mari-Jana 'M-J' Oboroceanu, 'American War and Military Operations Casualties: Lists and Statistics', Congressional Research Service, 26 February 2010, 3, http://www.fas.org/sgp/crs/natsec/RL32492.pdf, accessed June 2010.

18 George Bush, Remarks to the American Legislative Exchange Council, 1 March 1991, retrieved from John T. Woolley and Gerhard Peters, *The American Presidency Project* [online], Santa Barbara, Calif., http://www.presidency.ucsb.edu/ws/?pid=19351, accessed June 2010.

19 General H. Norman Schwarzkopf, with Peter Petre, *It Doesn't Take a Hero* (New York: Bantam Press, 1992), 418.

20 General Colin L. Powell, 'U.S. Forces: Challenges Ahead', *Foreign Affairs* 71 (1992): 36–41.

21 For an influential analysis along these lines see Andrew F. Krepinevich, 'The Military-Technical Revolution: A Preliminary Assessment', which was originally prepared for the Office of Net Assessment in 1992 (Washington, DC: Center for Strategic and Budgetary Assessments, 2002).

22 See, for example, Admiral Bill Owens, with Ed Offley, *Lifting the Fog of War* (New York: Farrar, Straus & Giroux, 2000).

23   Ibid., 14.

24   An important cautionary note in this regard was provided by Lawrence Freedman, 'The Revolution in Strategic Affairs', *Adelphi Paper 318* (Oxford: Oxford University Press for the International Institute of Strategic Studies, 1998).

25   'At Least: Slow the Slaughter', *New York Times*, 4 October 1992; Colin L. Powell, 'Why Generals Get Nervous', *New York Times*, 8 October 1992.

26   General Michael C. Short, 'Interview', http://www.pbs.org/wgbh/pages/ frontline/shows/kosovo/interviews/short.html, pp. 1, 3–4, accessed June 2010.

27   General Wesley K. Clark, *Waging Modern War: Bosnia, Kosovo and the Future of Combat* (Oxford: PublicAffairs, 2001), 203.

28   Lawrence Freedman, 'Victims and Victors: Reflections on the Kosovo War', *Review of International Studies* 26 (2000): 335–58.

29   Ibid., 357–8.

30   'Kosovo / Operation Allied Force After-Action Report', Report to Congress (unclassified), 31 January 2000, 10–12, http://www.dod.gov/ pubs/kaar02072000.pdf.

31   For a more convincing analysis see James Gow, *The Serbian Project and its Adversaries: A Strategy of War Crimes* (London: Hurst, 2003), 293–301.

32   Clark, *Waging Modern War*, 418–19.

Chapter 7

1   Although other interpretations are possible on this point. For example, John Stone, 'Al Qaeda, Deterrence, and Weapons of Mass Destruction', *Studies in Conflict and Terrorism* 32 (2009): 763–75.

2   Preface to President George W. Bush's introduction to *The National Security Strategy of the United States of America* (n.p.: September 2002), unpaginated.

3   George W. Bush, 'Address Before a Joint Session of the Congress on the United States Response to the Terrorist Attacks of September 2001', 20 September 2001, retrieved from John T. Wooley and Gerhard Peters, *The American Presidency Project* [online], Santa Barbara, Calif., http://www.presidency.ucsb.edu/ws/index.php?pid=64731&st=&st1=, accessed June 2010.

4   *Joint Vision 2020* (Washington DC: US Government Printing Office, 2000), 6.

5   *Military Transformation: A Strategic Approach* (Washington, DC: Department of Defense, 2003), 23.

6   Although Condoleezza Rice is not a Neoconservative per se, something of this teleological attitude is captured in comments she made shortly before becoming Bush's National Security Advisor in 2001. The 'United States and its allies are on the right side of history', she claimed. Likewise, 'As history marches towards markets and democracy, some states have

been left by the side of the road.' Condoleezza Rice, 'Promoting the National Interest', *Foreign Affairs* 79 (2000): 46, 60.

7   Bush publicly demanded that the Taliban hand over al-Qaeda's leadership in his address to Congress on 20 September 2001.

8   The tactical aspects of the early phase of the war are explored in Stephen Biddle, *Afghanistan and the Future of Warfare: Implications for Army and Defense Policy* (Carlisle, Pa.: Strategic Studies Institute, 2002).

9   Max Boot, 'The Struggle to Transform the Military', *Foreign Affairs* 84 (2005): 103–4.

10 Michael Gordon and Bernard Trainor, *Cobra II: The Inside Story of the Invasion and Occupation of Iraq* (London: Atlantic Books, 2006), 103.

11 There were also an additional 23,000 coalition troops in Iraq. The Brookings Institution, *Iraq Index: Tracking Variables Relevant to Reconstruction and Security in Post-Saddam Iraq* (updated 19 November 2003), http://www.brookings.edu/fp/saban/iraq/index20031119.pdf, accessed June 2010.

12 This account of early US operations in Iraq is based on Nigel Alwyn Foster, 'Changing the Army for Counterinsurgency Operations', *Military Review* 85 (2005): 2–15.

13 Major General Robert H. Scales, Jr, 'Culture-Centric Warfare', *US Naval Institute Proceedings* (October 2004): 33. See also Montgomery McFate, 'The Military Utility of Understanding Adversary Culture', *Joint Force Quarterly* 38 (2005): 42–8, who influentially argued that challenges such as the insurgency in Iraq would more readily yield to military initiatives that were informed by an understanding of the cultural milieu in which they were being conducted.

14 FM 3–24 / MCWP 3–33.5, *Counterinsurgency* (Washington, DC: Headquarters, Department of the Army, 2006), 3–6. The word 'culture' appears some 81 times in the manual.

15 Ibid., 1–27 notes that undue efforts to protect one's forces from attack can end up making them more vulnerable, in part because of the barriers they create to the collection of human intelligence. 'Ultimate success in COIN is gained by protecting the populace, not the COIN force. If military forces remain in their compounds, they lose touch with the people'. The force-protection issue is considered a salient paradox of counterinsurgency warfare, although there would seem to be no real paradox here per se. Rather, what we are looking at is an ironic reversal of outcomes.

16 Captain Nathan Finney, *Human Terrain Handbook* (Fort Leavenworth, Kans.: Human Terrain System, 2008).

17 I borrow the phrase 'war amongst the people' from General Sir Rupert Smith, *The Utility of Force: The Art of War in the Modern Age* (London: Allen Lane, 2005), passim.

18 Lieutenant General David H. Petraeus, 'Learning Counterinsurgency: Observations from Soldiering in Iraq', *Military Review* 86 (2006): 48.

19   'Stabilizing Iraq from the Bottom Up', Statement by Dr Stephen Biddle, Senior Fellow for Defense Policy, Council on Foreign Relations, Before the Committee on Foreign Relations, United States Senate, Second Session, 110th Congress, 2 April 2008, http://www.cfr.org/publication/15925/ stabilizing_iraq_from_the_bottom_up.html, accessed June 2010.

20   Stanley A. McChrystal, 'Commander's Initial Assessment', Unclassified (Kabul: Headquarters, International Security Assistance Force, 2009), 1–2, 2–4, 2–5.

21   'Operation Moshtarak', ISAF Joint Command News Release, Kabul, 13 February 2010, http://www.isaf.nato.int/images/stories/File/2010– 02-CA-059-Backgrounder-Operation%20Moshtarak.pdf, accessed June 2010; Barry Kolodkin, 'Operation Moshtarak in Afghanistan is Under Way', 13 February 2010, http://usforeignpolicy.about.com/b/2010/02/ 13/467.htm, accessed June 2010.

22   Tom Vanden Brook, 'Casualties caused by IEDs in Afghanistan on the rise', *USA Today*, 3 April 2009, http://www.usatoday.com/news/ military/2009–04-02-IEDs_N.htm, accessed June 2010. For details on the activities of the Joint Improvised Explosive Device Defeat Organization (JIEDDO) see its annual report for FY 2008 at https://www.jieddo. dod.mil/content/docs/20090625_FULL_2008_Annual_Report_ Unclassified_v4.pdf, accessed June 2010.

Bibliography

Andrew, Christopher. *The Defence of the Realm: The Authorized History of MI5*. London: Penguin, 2010

'At Least: Slow the Slaughter'. *New York Times*, 4 October 1992

Bassford, Christopher. *Clausewitz in English: The Reception of Clausewitz in Britain and America 1815–1945*. Oxford: Oxford University Press, 1994

Bell, David A. *The First Total War: Napoleon's Europe and the Birth of Modern Warfare*. London: Bloomsbury, 2007

Berlin, Isaiah, *The Sense of Reality: Studies in Ideas and their History*, ed. Henry Hardy (London: Chatto & Windus, 1996)

Betts, Richard K. 'Is Strategy an Illusion?' *International Security* 25 (2000): 5–50

Biddle, Stephen. *Afghanistan and the Future of Warfare: Implications for Army and Defense Policy*. Carlisle, Pa.: Strategic Studies Institute, 2002

—. 'Stabilizing Iraq from the Bottom Up', Statement by Dr Stephen Biddle, Senior Fellow for Defense Policy, Council on Foreign Relations, Before the Committee on Foreign Relations, United States Senate, Second Session, 110th Congress, 2 April 2008, http://www.cfr.org/publication/15925/stabilizing_iraq_from_the_bottom_up.html, accessed June 2010

Bismarck, Otto Prince von. *Bismarck: The Man and the Statesman*, trans. A.J. Butler. London: Smith, Elder & Co., 1898

—. *The Love Letters of Bismarck: Being Letters to His Fiancée and Wife, 1846–1889, Authorized by Prince Herbert von Bismarck and Translated from the German under the Supervision of Charlton T. Lewis*. New York: Harper Brothers, 1901

Bloch, I.S. *Is War Now Impossible? Being an Abridgement of the War of the Future in its Technical, Economic and Political Relations*. Aldershot: Gregg Revivals, in association with the Department of War Studies King's College London, 1991

Boot, Max. 'The Struggle to Transform the Military'. *Foreign Affairs* 84 (2005): 103–18

Bourrienne, M. de. *Memoirs of Napoleon Bonaparte, Vol I*. London: Richard Bentley, 1836

Brodie, Bernard. *The Absolute Weapon: Atomic Power and World Order*. New York: Harcourt Brace, 1946

—. 'The Development of Nuclear Strategy', *ACIS Working Paper No. 11*. Los Angeles, Calif.: Center for Arms Control and International Security, 1978

Brook, Tom Vanden. 'Casualties caused by IEDs in Afghanistan on the rise'. *USA Today*, 3 April 2009, http://www.usatoday.com/news/military/2009-04-02-IEDs_N.htm, accessed June 2010

The Brookings Institution. *Iraq Index: Tracking Variables Relevant to Reconstruction and Security in Post-Saddam Iraq* (updated 19 November 2003). http://www.brookings.edu/fp/saban/iraq/index20031119.pdf, accessed June 2010

Buckle, George Eliot (in succession to W.F. Monypenny) *The Life of Benjamin Disraeli, Earl of Beaconsfield, Volume V: 1868–1876*. London: John Murray, 1920

Bülow, Prince von, *Memoirs, Vol. II: 1903–1909*, trans. F.A. Voigt (London: Putnam, 1931)

Burke, Edmund. *The Writings and Speeches of Edmund Burke, Vol. IX: The Revolutionary War, 1794–1797*, ed. R.B. McDowall. Oxford: Clarendon, 1991

Bush, George. 'Remarks to the American Legislative Exchange Council, 1 March 1991', retrieved from John T. Woolley and Gerhard Peters, *The American Presidency Project* [online]. Santa Barbara, Calif., http://www.presidency.ucsb.edu/ws/?pid=19351, accessed June 2010

Bush, George W. 'Address Before a Joint Session of the Congress on the United States Response to the Terrorist Attacks of September 2001', 20 September 2001, retrieved from John T. Wooley and Gerhard Peters, *The American Presidency Project* [online]. Santa Barbara, Calif., http://www.presidency.ucsb.edu/ws/index.php?pid=64731&st=&st1=, accessed June 2010

Carnot, Lazare. *Correspondence générale de Carnot, Vol. IV: novembre 1793 – mars 1795*, ed. E. Charavay. Paris: Imprimerie Nationale, 1907

Clark, General Wesley K. *Waging Modern War: Bosnia, Kosovo and the Future of Combat*. Oxford: PublicAffairs, 2001

Clausewitz, Carl von. *On War*, trans. Colonel J.J. Graham (New York: Barnes & Noble, 2004)

Corvisier, André, ed. *A Dictionary of Military History*, trans. Chris Turner, rev. John Childs. Oxford: Basil Blackwell, 1994

Craig, Campbell. *Destroying the Village: Eisenhower and Thermonuclear War*. New York: Columbia University Press, 1998

Cronkite, Walter. *A Reporter's Life*. New York: Knopf, 1996

Dankbaar, Ben. 'Alternative Defense Policies and the Peace Movement'. *Journal of Peace Research* 21 (1984): 141–55

De Gaulle, General. *The Army of the Future*. London: Hutchinson, n.d. [1941]

Douhet, Giulio. *Command of the Air*, trans. Dino Ferrari. London: Faber & Faber, 1943

Duffy, Christopher. *The Military Experience in the Age of Reason*. London: Routledge & Keegan Paul, 1987

Dulles, John Foster. 'The Evolution of Foreign Policy'. *Department of State Bulletin*, 25 January 1954, 107–10

Earle, Edward Meade, ed. *Makers of Modern Strategy: Military Thought from Machiavelli to Hitler*. Princeton, N.J.: Princeton University Press, 1943

Eisenhower, Dwight D. 'The President's News Conference', 12 January 1955, retrieved from John T. Woolley and Gerhard Peters, *The American Presidency*

Project [online], Santa Barabara, Calif., http://www.presidency.ucsb.edu/ws/index.php?pid=10232, accessed June 2010

Enthoven, Alain C. and Wayne K. Smith. *How Much is Enough? Shaping the Defense Program 1961–1969*. New York: Harper & Row, 1971

Erickson, John. *The Road to Stalingrad: Stalin's War With Germany, Volume 1*. London: Weidenfeld & Nicolson, 1975

—. *The Road to Berlin: Stalin's War with Germany, Volume 2*. London: Weidenfeld & Nicolson, 1983

Esdaile, Charles J. *The Wars of Napoleon*. London: Longman, 1995

Etzold, Thomas H. and John Lewis Gaddis, eds. *Containment: Documents on American Policy and Strategy, 1945–1950*. New York: Columbia University Press, 1978

Finney, Captain Nathan. *Human Terrain Handbook*. Fort Leavenworth, Kans.: Human Terrain System, 2008

FM 3–24 / MCWP 3–33.5 *Counterinsurgency*. Washington, DC: Headquarters, Department of the Army, 2006

FM 17–100, *Armored Command Field Manual: The Armored Division*. Washington, DC: US Government Printing Office, 1944

FM 100–5 *Operations*. Washington, DC: Headquarters, Department of the Army, 1976

FM 100–5 *Operations*. Washington, DC: Department of the Army, 1986

Foster, Nigel Alwyn. 'Changing the Army for Counterinsurgency Operations'. *Military Review* 85 (2005): 2–15

Frederick the Great. *Instructions for his Generals*, trans. Brigadier General Thomas R. Phillips. Harrisburg, Penn.: Military Service Publishing Company, 1944

Freedman, Lawrence. *The Evolution of Nuclear Strategy*, 2nd edn. London: Macmillan in association with the International Institute for Strategic Studies, 1989

—. 'The Revolution in Strategic Affairs', *Adelphi Paper 318*. Oxford: Oxford University Press for the International Institute of Strategic Studies, 1998

—. 'Victims and victors: reflections on the Kosovo War'. *Review of International Studies* 26 (2000): 335–58

Freidberg, Aaron L. 'Why Didn't the United States Become a Garrison State?' *International Security* 16 (1992): 109–42

Fuller, Colonel J.F.C. *The Reformation of War*. London: Hutchinson, 1923

—. *The Foundations of the Science of War*. London: Hutchinson, 1926

—. *On Future Warfare*. London: Sifton Praed, 1928

Fuller, Major-General J.F.C. *Lectures on F.S.R. III (Operations Between Mechanized Forces)*. London: Sifton Praed, 1932

—. *Towards Armageddon: The Defence Problem and its Solution*. London: Lovat Dickson, 1937

Gates, David. *The Spanish Ulcer: A History of the Peninsular War*. London: George Allen & Unwin, 1986

Germains, Victor Wallace. *The 'Mechanization' of War.* London: Sifton Praed, 1927

Giap, Vo Nguyen. *The Military Art of People's War: Selected Writings of Vo Nguyen Giap*, ed. Russell Slater. New York: Monthly Review Press, 1970

Gordon, Michael and Bernard Trainor. *Cobra II: The Inside Story of the Invasion and Occupation of Iraq.* London: Atlantic Books, 2006

Gow, James. *The Serbian Project and its Adversaries: A Strategy of War Crimes.* London: Hurst, 2003

Gray, Colin S. 'Why Strategy Is Difficult'. *Joint Force Quarterly* 22 (1999): 6–12

Guderian, Major-General Heinz. *Achtung-Panzer! The Development of Armoured Forces, their Tactics and Operational Potential*, trans. Christopher Duffy. London: Arms & Armour Press, 1992

Halloran, Richard. 'Pentagon Draws up First Strategy for Fighting a Long Nuclear War'. *New York Times*, 30 May 1982

Henderson, Colonel G.F.R. *The Science of War: A Collection of Essays and Lectures 1892–1903*, ed. Captain Neill Malcolm. London: Longmans, Green & Co., 1905

Heuser, Beatrice. *Reading Clausewitz.* London: Pimlico, 2002

Horne, Alistair. *The Price of Glory: Verdun 1916.* London: Macmillan, 1962

Howard, Michael. *The Franco-Prussian War.* London: Rupert Hart-Davis, 1961

Ingrao, Charles. 'Paul W. Schroeder's Balance of Power: Stability or Anarchy?' *International History Review* 16 (1994): 681–700

Interview with General Michael C. Short, http://www.pbs.org/wgbh/pages/frontline/shows/kosovo/interviews/short.html, accessed June 2010

Ivashutin, Colonel General P., tr. Svetlana Savranskaya. 'Strategic Operations of the Nuclear Forces'. 28 August 1964. http://php.isn.ethz.ch/collections/colltopic.cfm?id=16248&lng=en, accessed May 2010

Joint Improvised Explosive Device Defeat Organization. Annual Report FY 2008. Https://www.jieddo.dod.mil/content/docs/20090625_FULL_2008_Annual_Report_Unclassified_v4.pdf, accessed June 2010

Joint Vision 2020. Washington DC: US Government Printing Office, 2000

Jomini, Baron Henri de. *Précis de l'Art de la Guerre.* Paris: n.p., 1838

Kahn, Herman. *On Escalation: Metaphors and Scenarios.* London: Pall Mall Press, 1965

Kaplan, Fred. *The Wizards of Armageddon.* New York: Simon & Schuster, 1983

Kaufmann, William W., ed. *Military Policy and National Security.* Princeton, N.J.: Princeton University Press, 1956

—. *The McNamara Strategy.* New York: Harper & Row, 1964

Kemp, Geoffrey, Robert L. Pfaltzgraff, Jr and Uri Ra'anan. *The Other Arms Race: New Technologies and Non-Nuclear Conflict.* Lexington, Mass.: Lexington Books, 1975

Kennedy, John F. 'Radio and Television Report to the American People on the Soviet Arms Buildup in Cuba', 22 October 1962, retrieved from the John F.

Kennedy Presidential Library and Museum, http://www.jfklibrary.org/Historical+Resources/Archives/Reference+Desk/Speeches/JFK/003POF03CubaCrisis10221962.htm, accessed May 2010

Kennedy, Paul, ed. *The War Plans of the Great Powers, 1880–1914*. London: George Allen & Unwin, 1979

King Jr, James E. 'Nuclear Plenty and Limited War'. *Foreign Affairs* 35 (1957): 238–56

Kissinger, Henry A. *Nuclear Weapons and Foreign Policy*. New York: Harper & Brothers, for the Council on Foreign Relations, 1957

Kolodkin, Barry. 'Operation Moshtarak in Afghanistan is Under Way', 13 February 2010, http://usforeignpolicy.about.com/b/2010/02/13/467.htm, accessed June 2010

'Kosovo / Operation Allied Force After-Action Report'. Report to Congress (unclassified), 31 January 2000, http://www.dod.gov/pubs/kaar02072000.pdf

Krepinevich, Andrew F. 'The Military-Technical Revolution: A Preliminary Assessment'. Washington, DC: Center for Strategic and Budgetary Assessments, 2002

Lasswell, Harold D. 'Sino-Japanese Crisis: The Garrison State versus the Civilian State'. *China Quarterly* (Shanghai) 2 (1937): 643–9

Leland, Anne and Mari-Jana 'M-J' Oboroceanu. 'American War and Military Operations Casualties: Lists and Statistics'. Congressional Research Service, 26 February 2010, http://www.fas.org/sgp/crs/natsec/RL32492.pdf, accessed June 2010

Liddell Hart, Captain B.H. *Paris, or the Future of War*. London: Kegan, Paul, Trench, Trubner, 1925

—. *When Britain Goes to War: Adaptability and Mobility*. London: Faber & Faber, 1935

—. *Europe in Arms*. London: Faber & Faber, 1937

—. *The Defence of Britain*. London: Faber & Faber, 1939

—. *The Revolution in Warfare*. London: Faber & Faber, 1946

Ludendorff, General. *The Nation at War*, trans. A.S. Rappoport. London: Hutchinson, n.d. [1936]

Luttwak, Edward N. 'The Operational Level of War'. *International Security* 5 (1980): 61–79

—. *Strategy: The Logic of War and Peace*. Cambridge, Mass.: Belknap, 1987

MacArthur, Douglas. *Reminiscences*. London: Heinemann, 1964

Mako, William P. *U.S. Ground Forces and the Defense of Central Europe*. Washington, DC: Brookings, 1983

Marshall, General George C. *The Winning of the War in Europe and the Pacific: Biennial Report of the Chief of Staff of the United States Army 1943 to 1945, to the Secretary of War*. New York: Simon & Schuster, 1945

McChrystal, Stanley A. 'Commander's Initial Assessment', Unclassified. Kabul: Headquarters, International Security Assistance Force, 2009

McFate, Montgomery. 'The Military Utility of Understanding Adversary Culture'. *Joint Force Quarterly* 38 (2005): 42–8

McNamara, Robert S. 'Remarks by Secretary McNamara, NATO Ministerial Meeting, 5 May 1962, Restricted Session' (Top Secret)

—, with Brian VanDeMark. *In Retrospect: The Tragedy and Lessons of Vietnam.* New York: Times Books, 1995

McNeill, William H. *The Pursuit of Power: Technology, Armed Force, and Society since A.D. 1000.* Oxford: Blackwell, 1983

McPherson, James M. *Battle Cry of Freedom: The American Civil War.* Oxford: Oxford University Press, 1988

Middlebrook, Martin. *The Schweinfurt-Regensburg Mission: American Raids on 17 August 1943.* London: Cassell, 2000

Miksche, F.O. *Blitzkrieg.* London: Faber & Faber, 1941

Military Transformation: A Strategic Approach. Washington, DC: Department of Defense, 2003

Millet, Allen R. and Williamson Murray, eds. *Military Effectiveness, Volume III: The Second World War.* Boston, Mass.: Allen & Unwin, 1988

Mitchell, William. *Winged Defense: The Development and Possibilities of Modern Airpower – Economic and Military.* New York: G.P. Putnam's Sons, 1925

Moltke, Helmuth von. *Essays, Speeches, and Memoirs of Field-Marshal Count Helmuth von Moltke, Volume II,* trans. Charles Flint McClumpha, Major C. Barter & Mary Herms. New York: Harper & Brothers, 1893

—. *Moltke's Military Correspondence 1870–71,* ed. Spencer Wilkinson. Aldershot: Gregg Revivals, 1991)

— *Moltke on the Art of War: Selected Writings,* trans. and ed. Daniel J. Hughes and Harry Bell. Novato, Calif.: Presidio, 1993

Morgenthau, Hans J. 'The Four Paradoxes of Nuclear Strategy'. *The American Political Science Review* 58 (1964): 23–35

—. 'We Are Deluding Ourselves in Vietnam'. *New York Times Magazine,* 18 April 1965

'National Security Decision Memorandum 242', 17 January 1974 (Top Secret / Sensitive)

The National Security Strategy of the United States of America (n.p.: September 2002)

Nixon, Richard. 'Third Annual Report to the Congress on Foreign Policy', 9 February 1972, retrieved from John T. Woolley and Gerhard Peters. *The American Presidency Project* [online]. Santa Barabara, Calif., http://www.presidency.ucsb.edu/ws/index.php?pid=3736&st=foreign&st1=policy, accessed May 2010

NSC-162/2, 'A Report to the National Security Council by the Executive Secretary on Basic National Security Policy', 30 October 1953

NSDD 12, 'Strategic Forces Modernization Program', 1 October 1981 (partial unclassified version)

'Operation Moshtarak', ISAF Joint Command News Release, Kabul, 13 February 2010, http://www.isaf.nato.int/images/stories/File/2010–02-CA-059-Backgrounder-Operation%20Moshtarak.pdf, accessed June 2010

Orwell, George. *The Collected Essays, Journalism and Letters of George Orwell, Volume II: My Country Right or Left, 1940–1943*, ed. Sonia Orwell and Ian Angus. London: Secker & Warburg, 1968

Osgood, Robert E. *Limited War Revisited*. Boulder, Col.: Westview Press, 1979

Overy, Richard. *Russia's War*. London: Allen Lane The Penguin Press, 1998

Owens, Admiral Bill, with Ed Offley. *Lifting the Fog of War*. New York: Farrar, Straus & Giroux, 2000

Paret, Peter, ed. *Makers of Modern Strategy: from Machiavelli to the Nuclear Age*. Oxford: Clarendon, 1986

Pentagon Papers: The Defense Department History of United States Decisionmaking on Vietnam, 4 vols, Gravel Edn. Boston, Mass.: Beacon Press, 1971–72 http://www.mtholyoke.edu/acad/intrel/pentagon/pent1.html, accessed June 2010

Pershing, John J. *My Experiences in the World War*. London: Hodder & Stoughton, 1931

Petraeus, Lieutenant General David H. 'Learning Counterinsurgency: Observations from Soldiering in Iraq'. *Military Review* 86 (2006): 45–55

Pflanze, Otto. *Bismarck and the Development of Germany: The Period of Unification, 1815–1871*. Princeton, N.J.: Princeton University Press, 1963

Phillips, Brig. Gen. Thomas R., trans. & ed. *Roots of Strategy: A Collection of Military Classics*. Mechanicsburg, Penn.: Stackpole, 1985

Powell, Colin L. 'Why Generals Get Nervous'. *New York Times*, 8 October 1992

Powell, General Colin L. 'U.S. Forces: Challenges Ahead'. *Foreign Affairs* 71 (1992): 32–45

Powell, Colin, with Joseph E. Persico. *A Soldier's Way: An Autobiography*. London: Hutchinson, 1995

Presidential Directive / NSC-59, 'Nuclear Weapons Employment Policy', 25 July 1980 (Secret with Top Secret attachment)

Rice, Condoleezza. 'Promoting the National Interest'. *Foreign Affairs* 79 (2000): 45–62

Ridgway, Matthew B. *The War in Korea: How We Met the Challenge, How All-Out Asian War Was Averted, Why MacArthur Was Dismissed, Why Today's War Objectives Must Be Limited*. London: Barrie & Rockliff, The Crescent Press, 1968

Ritter, Gerhard. *The Schlieffen Plan: Critique of a Myth*, trans. Andrew and Eva Wilson. London: Oswald Wolf, 1958

Roosevelt, Franklin D. 'Annual Message to Congress on the State of the Union', 6 January 1941, retrieved from John T. Woolley and Gerhard Peters, *The American Presidency Project* [online], Santa Barabara, Calif., http://www.presidency.ucsb.edu/ws/index.php?pid=16092, accessed April 2010

Rothenburg, Gunther. *The Art of Warfare in the Age of Napoleon*. London: Batsford, 1977

Savage, D.S., George Woodcock, Alex Comfort and George Orwell. 'Pacifism and the War: A Controversy'. *Partisan Review* 9 (1942): 414–21

Scales, Jr, Major General Robert H. 'Culture-Centric Warfare.' *US Naval Institute Proceedings* (October 2004): 32–6

Schelling, T. C. 'Controlled Response and Strategic Warfare', *Adelphi Paper 19*. London: IISS, 1965

Schelling, Thomas C. *The Strategy of Conflict*. Cambridge, Mass.: Harvard University Press, 1960

—. *Arms and Influence*. New Haven, Conn.: Yale University Press, 1966

Schlieffen, Alfred von. *Alfred von Schlieffen's Military Writings*, trans. and ed. Robert T. Foley (London: Frank Cass, 2003)

Schwarzkopf, General H. Norman, with Peter Petre. *It Doesn't Take a Hero*. New York: Bantam Press, 1992

Seversky, Alexander. *Victory Through Air Power*. New York: Simon & Schuster, 1942

Short, General Michael C. 'Interview'. Http://www.pbs.org/wgbh/pages/frontline/shows/kosovo/interviews/short.html, accessed June 2010

Sieyès, Emmanuel Joseph. *Qu'est-ce que le Tiers État?* Paris: Champs Flammarion, 1988

Simmons, Clifford, ed. *The Objectors*. Isle of Man: Anthony Gibbs & Phillips, n.d. [1965]

Simpkin, Richard, in association with John Erickson. *Deep Battle: The Brainchild of Marshall Tukhachevskii*. London: Brassey's, 1987

Smith, Munroe. 'Military Strategy Versus Diplomacy in Bismarck's Time and Afterward.' *Political Science Quarterly* 30 (1915): 37–81

Smith, General Sir Rupert. *The Utility of Force: The Art of War in the Modern Age*. London: Allen Lane, 2005

Speer, Albert. *Inside the Third Reich: Memoirs by Albert Speer*, trans. Richard and Clara Winston. London: Weidenfeld & Nicolson, 1970

Stone, John. *The Tank Debate: Armour and the Anglo-American Tradition*. Amsterdam: Harwood Academic Publishers, 2000

—. 'Al Qaeda, Deterrence, and Weapons of Mass Destruction.' *Studies in Conflict and Terrorism* 32 (2009): 763–75

Taylor, General Maxwell D. *The Uncertain Trumpet*. New York: Harper, 1959

Tocqeville, Alexis de. *De la démocratie en Amérique*. Paris: Garnier-Flammarion, 1981

Tuchman, Barbara. *The Guns of August*. New York: Macmillan, 1962

Tufte, Edward R. *The Visual Display of Quantitative Information*, 2nd edn. Cheshire, Conn.: Graphics Press, 2001

Tulard, Jean. *Napoleon: The Myth of the Saviour*, trans. Teresa Waugh. London: Weidenfeld Nicolson, 1984

United Nations General Assembly, A/1435, 'The Problem of the Independence of Korea', 7 October 1950

Webster, Charles and Noble Frankland. *The Strategic Air Offensive Against Germany 1939–1945*, 4 Vols. London: HMSO, 1961

Weinberger, Caspar W. 'The Uses of Military Power', Remarks Prepared for Delivery by the Hon. Caspar W. Weinberger, Secretary of Defense, to the National Press Club Washington, D.C., 28 November 1984. Http://www.pbs.org/wgbh/pages/frontline/shows/military/force/weinberger.html, accessed June 2010

Westmoreland, General W.C., USA, Commander, U.S. Military Assistance Command, Vietnam. *Report on Operations in South Vietnam January 1964 – June 1968*. Washington, DC: US Government Printing Office, n.d. [1968]

Westmoreland, General William C. *A Soldier Reports*. New York: Doubleday, 1976

Wohlstetter, Albert. 'The Delicate Balance of Terror', Rand Report P-1472. Santa Monica, Calif.: Rand, 1958

Wright, Quincy. *A Study of War*, 2nd edn. Chicago, Ill.: University of Chicago Press, 1965

X [George Kennan]. 'The Sources of Soviet Conduct.' *Foreign Affairs* 25 (1946–47): 566–82

Index

Lightning Source UK Ltd.
Milton Keynes UK
UKHW020049160222
398766UK00003B/85

9 781350 106246